Lecture Notes in Mathematics

Edited by A. Dold and B. Eckmann

824

D. Frank Hsu

Cyclic Neofields and Combinatorial Designs

Springer-Verlag
Berlin Heidelberg New York 1980

Author

D. Frank Hsu
Department of Mathematics
Fordham University
Bronx, NY 10458
USA

AMS Subject Classifications (1980): 05 B05, 05 B10, 05 B20, 05 C20, 12-02, 12 K99

ISBN 3-540-10243-4 Springer-Verlag Berlin Heidelberg New York
ISBN 0-387-10243-4 Springer-Verlag New York Heidelberg Berlin

Printing and binding: Beltz Offsetdruck, Hemsbach/Bergstr.
2141/3140-543210

Preface

The aim of this monograph is to present an account of the structure theory of cyclic neofields and to show their applications to various areas. An attempt has been made to keep this monograph as self-contained as possible.

A glance at the Table of Contents will reveal a little of what the monograph contains. Examples are given wherever necessary to help the readers absorb the theory and general constructions.

There is no formal prerequisite for the material, other than a graduate course in algebra covering non-associativity and a minimal acquaintance with combinatorics and number theory.

Some parts of this monograph are taken from my thesis written at the University of Michigan under the supervision of Professor Thomas F. Storer. I wish to thank him for introducing me to the wonderful world of cyclotomy theory and combinatorics, and for his encouragement and advice. Thanks are also due to Professors Donald J. Lewis, David Winter and Katta Murty who read an early draft of this monograph and made a number of useful suggestions.

TABLE OF CONTENTS

INTRODUCTION

The study of group difference sets can be regarded as going back to Kirkman [15], but was undertaken in more generality by Singer [22], and Bruck [2]. Given a difference set in a group, that difference set and all its translates in the group form a symmetric block design(see Hall [5]), and in fact, given any symmetric block design which has a sharply transitive collineation group, that block design can be shown to arise as the block design associated with a group difference set (see Bruck [2]).

The additive groups of finite fields provided a wealth of combinatorial designs (see Storer [23], and Singer [22]), and more recently, intensive generalizations of the finite field have been studied, in which either the additive or multiplicative group has been replaced by a loop (see Doner [4], Hughes [9], Knuth [16] or Paige [21]).

In the years around 1950, Bruck [3] and Paige [21] introduced and discussed the non-associative algebraic structures called "neofields". These are generalizations of fields in that the additive "commutative group" structure is relaxed to loops and hence the associativity of addition in the field is denied. Neofields are very different from skewfields where associativity of multiplication is denied.

Johnsen and Storer [10] have studied the concept of difference sets in loops and the block designs which arise as the collections of translates of such difference sets. Conversely, they show that an abstract block design (a square tactical configuration) can be regarded as a block design arising from a loop, and that if the loop has the right inverse property, a particular type of block design, called a principal block partial design, arises as the set of left translates of a right loop difference set.

In another part of the work [11], Johnsen and Storer consider the additive loop of a cyclic inverse property neofield (hereafter designated as CIP-neofield); a version of Lehmer's criterion [18] that the e-th powers of the generator of the neofield form a right loop difference set is shown to hold. A class of CIP-neofields of prime-power order is constructed, where each such CIP-neofield is obtained by altering the additive structure of the field of the same order. Briefly, they were able to "twist" the finite fields (additively) to produce many proper CIP-neofields of the prime-power order $p^{\alpha} \geq 11$. In the early 1970's, Doner [4] characterized the CIP-neofields, using an extension of the idea of "twisting" the fields. He has shown that for $v \equiv 0$, 6, 12, 15, 18, 21 (mod 24), and v = 10, no CIP-neofield of order v exists. For any other finite v, a construction of a CIP-neofield of order v is provided.

The existence of loop difference sets (and the related block design) in CIP-neofields as exhibited by Johnsen and Storer provided the motivation for the present work. In Chapter I of the present work, a necessary and sufficient condition for the existence of a general cyclic neofield N_v of order v is given in terms of an XMP-admissible partition of the residues in $Z_{v-1} - \{0\}$, when v is even and in $Z_{v-1} - \{(v-1)/2\}$ when v is odd. XMP-admissible partition is then lifted to an XIP-admissible partition and an LXP-admissible partition, and hence the admissible partition used by Doner [4] to construct CIP-neofields is clearly a special case of an XIP-admissible partition.

The results of Chapter I are used in Chapters III and IV to show that proper XMP-neofields, proper LXP-neofields (hence proper RXP- and CMP-neofields) exist for any finite order v, and proper XIP-neofields exist for any finite order v except $v \equiv 0 \pmod 6$ or $v = 10$. Moreover, we have shown (in Chapter II) that SIP-neofields (XIP-neofields with the property that $(x+y)(y+x) = xy$) can not exist for certain finite orders v (viz., $v \equiv 0, 1, 5, 6, 7, 11, 12, 13$ $15, 17, 18, 19, 21, 23 \pmod{24}$ or $v = 10$), and constructions for all remaining finite orders are provided. Also a product theorem for LXP-neofields of even order is given to provide a recursive construction of these neofields from lower order neofields.

Proper XIP-neofields of orders other than $v \equiv 0$ (mod 6) and $v = 10$ are constructed (Chapters II and III).

Proper LXP- and XMP-neofields of any finite order are constructed by combinatorial methods (Chapter IV). On the other hand, proper LXP-neofields of even order are constructed by using a number theory technique. It is also proved (Chapter II.2) that there does not exist any cyclic neofield N_v, $v > 4$ with both CIP and SIP.

It is mentioned in Bruck [3] that a completely symmetric loop gives rise to a Steiner triple system on the elements of the loop. Johnsen and Storer [13] have shown that the existence of a cyclic Steiner triple system of order v-1 is equivalent to the existence of a CIP-neofield of order v. Doner [4] has shown that the existence of an odd order CIP-neofield is equivalent to the existence of an "almost" cyclic Steiner triple system. In Chapter V, we show that XIP-, LXP- and XMP-neofields give rise to certain triple systems. Totally proper XIP-neofields give rise to BIBD's with $\lambda = 2$; totally proper LXP-neofields give rise to BIBD's with $\lambda = 3$; and totally proper XMP-neofields give rise to BIBD's with $\lambda = 6$. Also, a reconstruction theorem for proper XIP-neofield of even order is given (Chapter V.4). Hence the lattice relation of different-property cyclic neofields analogous to Galois theory makes sense (Chapter V.5). It is pratically assured that totally proper cyclic neofields of different property can be reconstructed from appropriate BIBD's with different values of λ.

It is well-known that triple systems with $\lambda = 2$, $\lambda = 3$ or $\lambda = 6$ can be constructed by repeating each block twice, three times or six times, respectively in a Steiner triple system. But the above BIBD's do have much better properties, since each of these BIBD's consists of "almost" distinct blocks.

In Chapter VI, it is proved that the existence of a cyclic neofield N_v of order v, the existence of an I-matrix A_{v-2} of order v-2, the existence of an N-permutation π of Z^*_{v-1} and the existence of an N-graph G_{v-2} of order v-2 are equivalent to each other, It is also shown that for a prime number 2n+1, the matrix J^*_{2n} ($a_{ii} = 0$, $a_{ij} = 1$, $\forall i \neq j$) can be decomposed into the sum of 2n-1 I-matrices corresponding to cyclic neofields of different properties. Moreover, the above characterization of cyclic neofields gives the connection between the study of cyclic neofield, finite projective planes, permutation groups, matrix theory and statistical designs. This diversity of applications of cyclic neofields makes us recognize similarities among problems from different subjects of application.

Some outstanding problems have been mentioned in the last section of the work, and several of these are accessible to (a developed version of) the techniques introduced herein. In Appendix I, we list a construction of type a) XIP-admissible partitions. This is due to Doner in a slightly different form, but he used this only to construct CIP-neofields. In Appendix II, we list a Fortran program to obtain all cyclic neofields of every finite order. Cyclic neofields of order less than eleven are also given.

CHAPTER I

ADDITIVE STRUCTURE IN CYCLIC NEOFIELDS

Sec. 1. Preliminary Definitions and Results

Definition I.1: $\langle S, +\rangle$, a set S together with a binary operation +, is a loop, provided

a) there exists a $0 \in S$ such that

$$\forall a \in S: a+0 = 0+a = a;$$

b) for any choice of $a,b \in S$, the equations

$$\begin{cases} x+a = b \\ a+y = b \end{cases}$$

have unique solutions $x,y \in S$.

(Equivalently, $\langle S,+\rangle$ is a loop provided its addition table can be written as a normalized Latin square.)

Definition I.2: $\langle S,+,\cdot\rangle$, a set S with two binary operations + and $\cdot$, is a neofield provided

a) $\langle S,+\rangle$ is a loop;

b) $\langle S-\{0\},\cdot\rangle$ is a group;

c) multiplication distributes over addition on both the left and right.

Definition I.3: A neofield $\langle S,+,\cdot\rangle$ is commutative if addition is commutative. A neofield $\langle S,+,\cdot\rangle$ has the right (left) inverse property provided $\forall a\in S, \exists (-a)_R \in S$ $(\exists (-a)_L \in S)$ such that $\forall x\in S$, $(x+a)+(-a)_R = x$ $((-a)_L+(a+x) = x)$ and $\langle S,+,\cdot\rangle$ has inverse property provided it has both the right and left inverse property.

Definition I.4: A neofield $\langle S,+,\cdot\rangle$ has the exchange-inverse-property provided $\forall a\in S, \exists (-a)_L \in S$ such that $\forall x\in S$, $(-a)_L+(x+a) = x$.

Definition I.5: A neofield $\langle S,+,\cdot\rangle$ is cyclic if $\langle S-\{0\},\cdot\rangle$ is a cyclic group.

We shall be interested in cyclic neofields of finite order. Hereafter, $N_v = \langle S,+,\cdot\rangle$ shall be a special notation for a finite cyclic neofield of order v, with multiplicative group $\langle S-\{0\},\cdot\rangle$ given in terms of a generator a:

$$S-\{0\} = \{1,a,a^2,\cdots,a^{v-2}\}.$$

Hence the multiplicative structure in N_v is given implicitly in our notation, and we shall pursue the characterization of addition in terms of this notation.

Lemma I.6*: In a cyclic neofield N_v with generator a, we have

* This Lemma is due to Paige [21].

$$(-1)_R = \begin{cases} 1 \text{ for } v \text{ even}, \\ a^{(v-1)/2} \text{ for } v \text{ odd}. \end{cases}$$

Proof: If v is even, $-1 \neq 0$, so $-1 = a^h$ for some $h \in \{0,1,2,\ldots,v-1\}$. But then

$$a^{2h} = (a^h)(a^h) = (-1)(-1) = 1,$$

i.e., $2h = 0$ or $v-1$. Since v is even, $2h = 0$ and $-1 = a^o = 1$

If v is odd, let p(S) be the product of the v-1 distinct elements of $S-\{0\}$. Let $T(x) = 1+x$, i.e., T(x) is the entry under column x on the second row of the addition table. The equation $T(x) = b\cdot x$ has a solution for all $b \neq 1$ of $S-\{0\}$. For if $T(x)/x = T(y)/y$ for $x \neq y$, let $x = a^r$, $y = a^s$. We have $a^s(1+a^r) = a^r(1+a^s)$, i.e., $a^s+a^{r+s} = a^r + a^{r+s}$. Then $a^r = a^s$, hence $r \equiv s \pmod{v-1}$, i.e., $x = y$. This is a contradiction. Moreover, let h be such that $T(h) = 0\cdot h = 0$. Taking the product of all these equations, we have

$$\pi T(x) = (\pi b)\cdot(\pi x),$$

where x ranges over all elements of $S-\{0\}$ except h and b ranges over all elements of $S-\{0\}$ except 1. Hence $\pi T(x) = p(S)$, $\pi b = p(S)$ and $\pi x = p(S)\cdot h^{-1}$. Then we have $p(S) = p(S)\cdot p(S)\cdot h^{-1}$, i.e.,

$$a^{(v-1)/2} = a^{(v-1)/2} \cdot a^{(v-1)/2} \cdot h^{-1},$$

since $p(S) = 1\cdot a\cdot a^2\cdots a^{v-2} = a^{(v-1)/2}$, when v is odd. That is: $h^{-1} = a^{(v-1)/2}$, i.e., $h = a^{(v-1)/2}$. Therefore $1+h = T(h) = 0$, when $h = a^{(v-1)/2}$. Since the addition table is a Latin square, $a^{(v-1)/2}$ is the only element such that $1+a^{(v-1)/2} = 0$, i.e., $-1 = a^{(v-1)/2}$ for v odd. q.e.d.

From the above lemma, we have $(-1)_R = (-1)_L$. Hence for every x in a cyclic neofield, we have

$$[x+(-1)_R x] = [1+(-1)_R]x = 0\cdot x = 0$$

and

$$[(-1)_L x+x] = [(-1)_L +1]x = 0\cdot x = 0.$$

Therefore $(-1)_R x = (-1)_L x = -x$ is the unique two-sided negative of x. We write $(-x)_R = (-x)_L = -x$.

Note that since multiplication is distributive over addition, we have complete information about the addition in N_v if we know each of these sums

$$T(a^k) = 1+a^k = a^n, \quad k\in\{0,1,2,3,\ldots,v-2\}.$$

This is called the presentation function of N_v.

Since every cyclic neofield of finite order has the exchange-minus-property: $-(x+y) = -x-y$, we call them XMP-neofields. A cyclic neofield with right (left) inverse property is called an RXP(LXP)-neofield. A cyclic neofield with commutative property is called a CMP-neofield, and with exchange-inverse-property is called an XIP-neofield. A cyclic neofield with both RXP and LXP is called a CIP-neofield.

Now we define some mappings on the family of all cyclic neofields.

Definition I.7: The mappings CD, RD, T are defined as follows: for any cyclic neofield S, and $x+y = z$ in S, the additions in CD(S), RD(S) and T(S) are defined as follows:

a) $(-x)+z = y$ in CD(S); b) $z+(-y) = x$ in RD(S);

c) $y+x = z$ in T(S).

Lemma I.8: Transformations RD, CD, and T have the following properties.

i) $RD^2 = CD^2 = T^2 = I$, the identity mapping;

ii) $RD\cdot CD\cdot RD\cdot T = I$;

iii) $CD\cdot RD\cdot CD\cdot T = I$;

iv) $(CD\cdot RD)^3 = I$;

v) $CD\cdot RD\cdot CD = T$;

vi) $RD\cdot CD\cdot RD = T$.

Proof: We only prove ii). The remaining parts are similar. Let $x+y = z$ in a cyclic neofield S. Then we have $y+x = z$ in T(S). Hence we have $z+(-x) = y$ in $(RD\cdot T)(S)$ and $(-z)+y = -x$ in $(CD\cdot RD\cdot T)(S)$. Therefore we have $(-x)+(-y) = (-z)$, i.e., $x+y = z$ in $(RD\cdot CD\cdot RD\cdot T)(S)$. It follows that $RD\cdot CD\cdot RD\cdot T = I$. q.e.d.

Lemma I.9: For a cyclic neofield S with XMP; RD(S), CD(S), T(S) are also cyclic neofields.

Proof: That T(S) is a cyclic neofield is clear. We only show that CD(S) is a cyclic neofield. Recall that if $x+y = z$ in S, then $(-x)+z = y$ in CD(S). Let z-column be any column on CD(S). We claim that all the entries on the z-column are distinct. For if the entries $(x_1,z) = (x_2,z)$ in CD(S), say $(x_1,z) = y_1 = y_2 = (x_2,z)$. Then we have

$$x_1+z = y_1 \text{ and } x_2+z = y_2 \text{ in CD(S).}$$

Hence in S = CD(CD(S)), we have

$$(-x_1)+y_1 = z = (-x_2)+y_2.$$

Since $y_1 = y_2$ and the addition table of S is a Latin square, we have $x_1 = x_2$. Next we claim that all the entries on an arbitrary row of CD(S) are distinct. For if entries $(x,z_1) = (x,z_2) = y$, then $x+z_1 = x+z_2 = y$ in CD(S). Hence $(-x)+y = z_1$ and $(-x)+y = z_2$ in S. Thus $z_1 = z_2$. Therefore the addition table of CD(S) is a Latin square. The CD transformation doesn't change the multiplication. This completes the proof. q.e.d.

Lemma I.10: For a cyclic neofield S with RXP, RD(S)=S with LXP, CD(S)=S; with CMP, T(S)=S. Moreover, RD(S) has LXP and CD(S) has RXP if S has CMP.

Proof: We only prove the first part of the lemma. The proof of the remaining parts are similar. A cyclic neofield S with RXP has $x+y = z$ and $(x+y)+(-y) = x$, i.e., $z+(-y) = x$. Hence in RD(S), we have $z+(-y) = x$ and $x+(+y) = z$, i.e.,

RD = I. *q.e.d.*

Note that the converse of the first part of the lemma is also true. For example, if RD(S) = S for a cyclic neofield S, then S has RXP.

Lemma I.11: For an XIP-neofield S, RD(S)= CD(S)= T(S).

Proof: In the XIP-neofield S, we have (-y)+(x+y) = x, i.e., we have x+y = z and (-y)+z=x. Hence in CD(S), we have (+y)+x = z. Then in (T·CD)(S), we have x+y = z. It follows that T·CD = I, i.e., CD = T. Then by Lemma I.8, RD = CD = T. *q.e.d.*

Corollary I.12: For an XMP-neofield, (-y)+(x+y) = x if and only if ((-y)+x)+y = x.

Lemma I.13: For a CIP-neofield S,RD(S)= CD(S)= S.

Proof: Since CIP-neofield has RXP and LXP, we have RD = CD = I by Lemma I.10. *q.e.d.*

Lemma I.14: A finite CIP-neofield is commutative. Hence in a CIP-neofield, we have RD = CD = T = I.

Proof: Let S be a finite CIP-neofield. Then S has RXP and LXP, i.e., RD = CD = I. Hence

$$\underset{\text{in S}}{x+y = z} \overset{\text{RD}}{\Longrightarrow} \underset{\text{in S}}{z+(-y) = x} \overset{\text{CD}}{\Longrightarrow} \underset{\text{in S}}{(-z)+x = -y}$$

$$\overset{\text{RD}}{\Longrightarrow} (-y)+(-x) = -z \text{ in S},$$

i.e., we have y+x = z in S. Hence S is commutative. Combining these with Lemma I.13, we have RD = CD = T = I.

q.e.d.

Summarizing the above lemmas for RD, CD and T transformations, we have the following pictures:

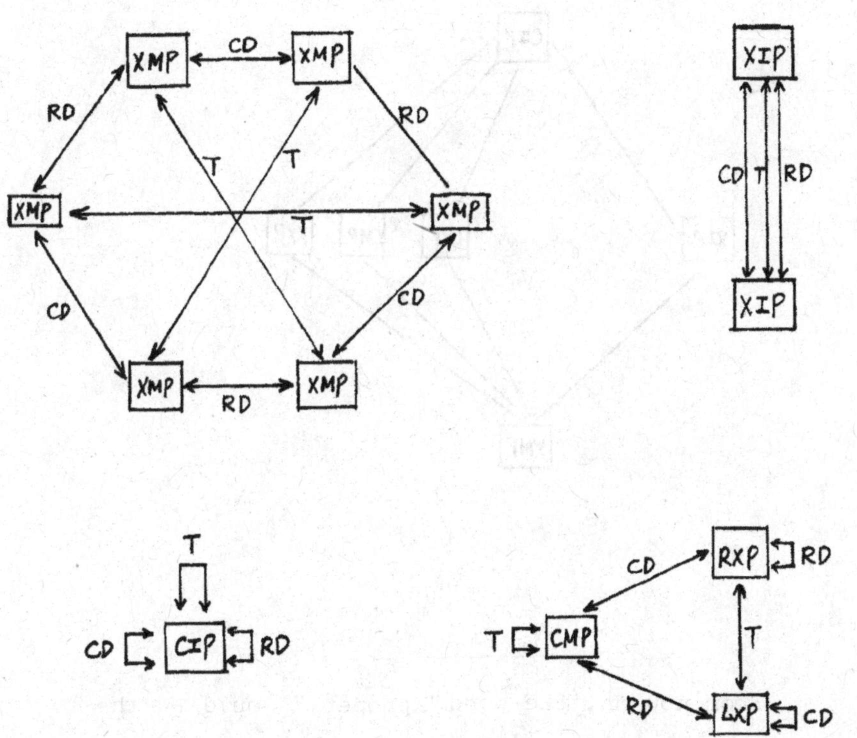

Figure 1

Note that the converse of Lemma I.14 is false: A finite commutative cyclic neofield need not be CIP. Hence we have the following lattice of cyclic neofields with different properties. One neofield sits above the other if it has stronger algebraic properties. A finite field is a finite CIP-neofield with associative addition.

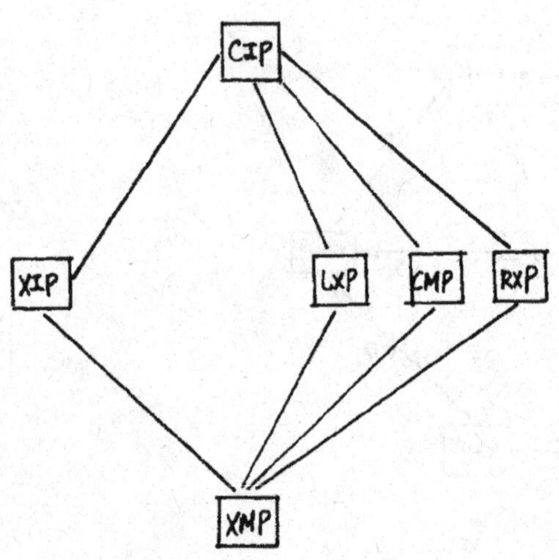

Figure 2

From now on, the word " proper " would be the adjective to any class of cyclic neofield (in the above lattice) which does not have a stronger property. Throughout this note, $Z_n^* = Z_n - \{0\}$, if n is odd; $Z_n^* = Z_n - \{\ell\}$, if n is even and $\ell = n/2$.

In the following two lemmas, we give some properties which a general XMP-neofield does not possess.

Lemma I.15: For N_v, an XIP-neofield of order v, with generator a,

a) if v is even, then any of the statements

i) $1+a^k = a^n$,

ii) $1+a^{n-k} = a^{-k}$,

iii) $1+a^{-n} = a^{k-n}$,

implies the remaining two;

b) if v is odd, let $\ell = \frac{v-1}{2}$; then any of the statements

i) $1+a^k = a^{n+\ell}$,

ii) $1+a^{n-k} = a^{-k+\ell}$,

iii) $1+a^{-n} = a^{k-n+\ell}$,

implies the remaining two.

Proof: (a) v even,

i) $\Rightarrow$ ii);

$$1+a^k = a^n \Rightarrow a^k+(1+a^k) = a^k+a^n = 1$$
$$\Rightarrow 1+a^{n-k} = a^{-k}.$$

ii) $\Rightarrow$ iii);

$$1+a^{n-k} = a^{-k} \Rightarrow a^{n-k}+(1+a^{n-k}) = a^{n-k}+a^{-k} = 1$$
$$\Rightarrow 1+a^{-n} = a^{k-n}.$$

iii) $\Rightarrow$ i);

$$1+a^{-n} = a^{k-n} \Rightarrow a^{-n}+(1+a^{-n}) = a^{-n}+a^{k-n} = 1$$
$$\Rightarrow 1+a^k = a^n.$$

(b) v odd,

i)$\Rightarrow$ii);

$$1+a^k = a^{n+\ell} \Rightarrow (-a^k)+(1+a^k) = (-a^k)+a^{n+\ell} = 1$$
$$\Rightarrow a^k+(-a^{n+\ell}) = -1$$
$$\Rightarrow a^k+a^{\ell}\cdot a^{n+\ell} = -1$$
$$\Rightarrow a^k+a^{n+2\ell} = -1$$
$$\Rightarrow a^k+a^n = a^{\ell}$$
$$\Rightarrow 1+a^{n-k} = a^{\ell-k} = a^{-k+\ell}.$$

ii)$\Rightarrow$iii);

$$1+a^{n-k} = a^{-k+\ell} \Rightarrow (-a^{n-k})+(1+a^{n-k}) = -a^{n-k}+a^{-k+\ell}$$
$$= 1$$
$$\Rightarrow a^{n-k+\ell}+a^{-k+\ell} = 1$$
$$\Rightarrow 1+a^{-n} = a^{k-n-\ell} = a^{k-n+\ell},$$

iii)$\Rightarrow$i);

$$1+a^{-n} = a^{k-n+\ell} \Rightarrow (-a^{-n})+(1+a^{-n}) = (-a^{-n})+a^{k-n+\ell}$$
$$= 1$$
$$\Rightarrow a^{-n+\ell}+a^{k-n+\ell} = 1$$
$$\Rightarrow 1+a^k = a^{n-\ell} = a^{n+\ell}.$$

q.e.d.

Lemma I.16: For N_v, an LXP-neofield of order v, with generator a,

a) if v is even, then the following two statements are equivalent,

i) $1+a^k = a^n$,

ii) $1+a^n = a^k$;

b) if v is odd, then the following two statements are equivalent (where $\ell = \frac{v-1}{2}$),

i) $1+a^k = a^{n+\ell}$,

ii) $1+a^n = a^{k+\ell}$.

<u>Proof</u>: (a) v even,

i) $\Rightarrow$ ii);

$$1+a^k = a^n \Rightarrow 1+(1+a^k) = 1+a^n = a^k$$
$$\Rightarrow 1+a^n = a^k.$$

ii) $\Rightarrow$ i);

$$1+a^n = a^k \Rightarrow 1+(1+a^n) = 1+a^k = a^n.$$

(b) v odd,

i) $\Rightarrow$ ii);

$$1+a^k = a^{n+\ell} \Rightarrow (-1)+(1+a^k) = (-1)+a^{n+\ell} = a^k$$
$$\Rightarrow a^{\ell}+a^{n+\ell} = a^k$$
$$\Rightarrow 1+a^n = a^{k-\ell} = a^{k+\ell}.$$

ii) $\Rightarrow$ i);

$$1+a^n = a^{k+\ell} \Rightarrow (-1)+(1+a^n) = (-1)+a^{k+\ell} = a^n$$
$$\Rightarrow a^{\ell}+a^{k+\ell} = a^n$$
$$\Rightarrow 1+a^k = a^{n-\ell} = a^{n+\ell}.$$

<u>q.e.d.</u>

The arguments similar to the previous two lemmas for RXP and CMP are omitted here, since RXP and CMP are related to LXP by T and RD in Figure 1. Since a cyclic neofield with both XIP and LXP would be a CIP-neofield, the argument for CIP-neofields is the combination of Lemmas I.15 and I.16. CIP-neofields are discussed in Doner[4]. But it will become clear later that CIP-neofields are special cases of either XIP-neofields or LXP-neofields.

Sec 2. XMP- and LXP-Admissible Partitions of $\mathbb{Z}_{v-1}$

Lemma I.17: Let N_v be an XMP-neofield of order v. Then

a) if v is even, $1+a^k = a^n$ in N_v implies that each of the conditions:

i) $k \not\equiv 0 \pmod{v-1}$

ii) $n \not\equiv 0 \pmod{v-1}$

iii) $k \not\equiv n \pmod{v-1}$

holds for the pair (k,n)

b) if v is odd, $1+a^k = a^{n+\ell}$ in N_v (where $\ell = \frac{v-1}{2}$)

implies that each of the conditions

i) $k \not\equiv \ell \pmod{v-1}$

ii) $n \not\equiv \ell \pmod{v-1}$

iii) $k \not\equiv n+\ell \pmod{v-1}$

holds for the pair (k,n).

<u>Proof</u>: In either a) or b), the condition i) excludes -1 from the role of a^k. This is obvious since, by Lemma I.14, $-1 = 1$ for v even and $-1 = a^\ell$ for v odd, then $1+(-1) = 0$ is fixed. The condition ii) guarantees that all elements in the second row of the addition table are distinct from 1 except $1+0 = 1$. The condition iii) simply says that the presentation function is a derangement*, otherwise the addition table would not be a Latin square. <u>q.e.d.</u>

<u>Definition I.18</u>: Call a pair (k,n) <u>admissible (mod v-1)</u> if it satisfies each of a) i), a) ii), and a) iii) when v is even ane each of b) i), b) ii), and b) iii) when v is odd in the above lemma.

* See pages 9 and 24 in Hall[5].

Lemma I.19: Let N_v be any XMP-neofield of even order v. Let $1+a^k = a^n$ and $1+a^s = a^t$ be any two relations in N_v. Then $|\{k,n\}| = |\{s,t\}| = 2$.

Proof: It follows immediately from Lemma I.17. q.e.d.

Lemma I.20: Let N_v be any XMP-neofield of odd order v. Let $1+a^k = a^{n+\ell}$ and $1+a^s = a^{t+\ell}$ be any two relations in N_v, where $\ell = \frac{v-1}{2}$. Then $|\{k,n\}| = |\{s,t\}| = 2$ with one exception: there exists a unique number $q \in \mathbb{Z}_{v-1} - \{\ell\}$ such that $1+a^q = a^{q+\ell}$.

Proof: Let $T(x) = 1+x$ be the presentation function of N_v. In the proof of Lemma I.16, we have

$$T(x)/x \neq T(y)/y \text{, if } x \neq y \text{ and } x,y \neq 0,a^\ell.$$

With regard to the function T(x), x runs over all elements in $\{0,1,a,a^2,\ldots a^{v-2}\} - \{0,a^\ell\}$ and T(x) runs over all elements in $\{0,1,a,a^2,\ldots a^{v-2}\} - \{0,1\}$. By the above observation, (T(x)/x)'s are distinct. Hence T(x)/x runs over all elements in $\{a,a^2, a^3,\ldots a^{v-2}\}$. Hence

$$\left|\bigcup_{\substack{x\neq 0\\ x\neq a^\ell}} \left(T(x)/x\right)\right| = \left|\{a,a^2,a^3,\ldots a^{v-2}\}\right| = v-2.$$

Therefore there must exist some x such that $T(x)/x = a^\ell$. Let $x = a^q$ (of course $q \neq \ell$ and $x \neq 0$). We then have $1+a^q = a^{q+\ell}$.

The remaining part of the lemma is trivial. q.e.d.

Definition I.21: For v even, a collection $X = \{(k_1,n_1), (k_2,n_2),\ldots,(k_{v-2}, n_{v-2})\}$ of ordered pairs of $\mathbf{Z}_{v-1}-\{0\}$ is said to be an XMP-admissible partition of $\mathbf{Z}_{v-1}-\{0\}$ provided each pair (k_i,n_i) is admissible (mod v-1) in the sense of Definition I.18 and

i) $\bigcup_{i=1}^{v-2}\{k_i,n_i\}$ consists of each element of $\mathbf{Z}_{v-1}-\{0\}$ exactly twice, each element of $\mathbf{Z}_{v-1}-\{0\}$ appears in two pairs of X, once as the first element, once as the second element,

ii) $\bigcup_{(k_i,n_i)\in X}\{n_i-k_i\} = \mathbf{Z}_{v-1}-\{0\}$.

For v odd, a collection $X = \{(k_1,n_1),\ldots,(k_{v-3},n_{v-3}), (k_{v-2}, k_{v-2})\}$ of ordered pairs of $\mathbf{Z}_{v-1}-\{\ell\}$ is said to be an XMP-admissible partition of $\mathbf{Z}_{v-1}-\{\ell\}$ provided each pair (k_i,n_i) is admissible (mod v-1) in the sense of Definition I.18 and

i) each element of $\mathbf{Z}_{v-1}-\{\ell\}$ appears in exactly two pairs of X, once as the first element and once as the second element, except for one element which contributes the pair (k_{v-2}, k_{v-2}),

ii) $\bigcup_{i=1}^{v-3}\{n_i-k_i\}\cup\{0\} = \mathbf{Z}_{v-1}-\{\ell\}$.

Lemma I.22: For v even and N_v an XMP-neofield of order v, then $\{(k,n)\,|\,1+a^k = a^n\}$ is an XMP-admissible partition of $\mathbf{Z}_{v-1}-\{0\}$. For v odd and N_v an XMP-neofield of order v, let $\ell = \frac{v-1}{2}$; then $\{(k,n)\,|\,1+a^k = a^{n+\ell}\}$ is an XMP-admissible partition of $\mathbb{Z}_{v-1}-\{\ell\}$.

Proof: It follows from Lemma I.17 that in either case, (k,n) in $\{(k,n) \mid 1+a^k = a^n\}$ or in $\{(k,n) \mid 1+a^k = a^{n+\ell}\}$ is admissible (mod v-1). We only show i) and ii) of Definition I.21 for the case when v is even.

For every element in $\mathbb{Z}_{v-1}-\{0\}$, say k, we have $1+a^k = a^s$ and $1+a^t = a^k$. Hence (k,s) and (t,k) are in $\{(k,n) \mid 1+a^k = a^n\}$. Since the additive table is a Latin square, this proves i). To show ii), we note that $T(x)/x \neq T(y)/y$ with $x \neq y$ for $x, y \neq 0, 1$. For if this is not true, then $yT(x) = xT(y)$. Let $x = a^p$, $y = a^q$. Then $a^q(1+a^p) = a^p(1+a^q)$. Hence $a^q+a^{p+q} = a^p+a^{p+q}$. It follows that $p \equiv q$ (mod v-1), a contradiction. Therefore, we have

$$\left| \bigcup_{\substack{x \neq 0 \\ x \neq 1}} \left\{\frac{T(x)}{x}\right\} \right| = \left| \{a, a^2, a^3 \ldots, a^{v-2}\} \right| = v-2,$$

i.e., $\bigcup_{1+a^{k_i} = a^{n_i}} \{n_i - k_i\} = \mathbb{Z}_{v-1}-\{0\}$. This proves ii).

q.e.d.

The above Lemma indicates that a necessary condition that an XMP-neofield of order v exists is that an XMP-admissible partition of $\mathbb{Z}_{v-1}-\{0\}$ when v is even, and of $\mathbb{Z}_{v-1}-\{\ell\}$ when v is odd, exists. We will show that this is also sufficient and give some examples.

Theorem I.23: For $v \in \mathbb{Z}$, $v > 0$, let $G = \{1, a, a^2, \ldots a^{v-2}\}$ be the cyclic group of order v-1 and let $S = G \cup \{0\}$, with multiplication in S the extension of multiplication in G given by

$0 \cdot g = g \cdot 0 = 0$ for all $g \in S$. Suppose there exists an XMP-admissible partition XM:

$$\begin{cases} XMA = \{(k_1,n_1),\dots,(k_{v-2},n_{v-2})\}, \text{ if } v \text{ is even,} \\ XMB = \{(k_1,n_1),\dots,(k_{v-3},n_{v-3}),(K_{v-2},k_{v-2})\}, \text{ if } v \text{ is odd,} \end{cases}$$

and define an addition in S by (here $\ell = (v-1)/2$)

$$\begin{cases} 0+x = x+0 = x, \forall x \in S \\ 1+1 = 0, \text{ if } v \text{ is even; } 1+a^{\ell} = 0 \text{ if } v \text{ is odd} \\ 1+a^k = \begin{cases} a^n \text{ if } (k,n) \text{ is in XMA when } v \text{ is even} \\ a^{n+\ell} \text{ if } (k,n) \text{ is in XMB when } v \text{ is odd} \end{cases} \\ a^r+a^s = a^r(1+s^{s-r}), \forall r,s \neq 0 \end{cases}$$

Then the addition is well-defined and $N_v = \langle S,+,\cdot\rangle$ in an XMP-neofield of order v.

Proof: For v even, since every (k,n) in XMA is admissible, the presentation function $T(x) = 1+x$ with $x \neq 0, 1$, is a derangement. By Definition I.21, $\bigcup_{i=1}^{v-2} \{n_i \mid (k_i,n_i) \in XMA\} = \mathbb{Z}_{v-1}-\{0\}$. Hence T(x)'s are distinct. Moreover, by Definition I.21, ii), we have $n_i-k_i \neq n_j-k_j$ if $i \neq j$. This implies that $T(x)/x \neq T(y)/y$ if $x \neq y$. Hence by the distributive hypothesis $a^r+a^s = a^r(1+a^{s-r})$, $r,s \neq 0$, the addition table is a normalized Latin square. Hence $N_v = \langle S,+,\cdot\rangle$ is an XMP-neofield.

For v odd, the proof is similar, except that $(\bigcup_{i=1}^{v-3}\{n_i \mid (k_i n_i) \in XMB\}) \cup \{k_{v-2}\} = \mathbb{Z}_{v-1}-\{\ell\}$, where $\ell = \frac{v-1}{2}$, and the differences of the exponents of T(x) and x run over $\mathbb{Z}_{v-1}-\{\ell\}$ by ii) of Definition I.21, when v is odd. q.e.d.

We now give two examples to illustrate how to construct XMP-neofield N_v from certain XMP-admissible partitions of $\mathbb{Z}_{v-1}-\{0\}$, when v is even and of $\mathbb{Z}_{v-1}-\{\ell\}$ when v is odd. The admissible partitions given in Chapter I are obtained from "trial and error" method. But we will give systematic constructions in Chapters II, III, and IV.

Example I.24: Let v = 9 and $\ell = 4$. We have the following XMP-admissible partition of $\mathbb{Z}_8-\{4\}$:

$$\text{XMB} = \{(0,6),(1,0),\ (2,3),\ (3,5),\ (5,2),\ (6,1),\ (7,7)\}.$$

Hence we have the XMP-neofield of order 9.

x	0	1	a	a^2	a^3	a^4	a^5	a^6	a^7
T(x)	1	a^2	a^4	a^7	a	0	a^6	a^5	a^3

Example I.25: Let v = 12. We have an XMP-admissible partition of $\mathbb{Z}_{11}-\{0\}$:

$$\text{XMA} = \{(1,5),\ (5,2),\ (2,4),\ (4,3),\ (3,8),\ (8,9),\ (9,7),\ (7,10,\ (10,6),\ (6,1)\}.$$

Hence we have the presentation function of this XMP-neofield:

x	0	1	a	a^2	a^3	a^4	a^5	a^6	a^7	a^8	a^9	a^{10}
T(X)	1	0	a^5	a^4	a^8	a^3	a^2	a	a^{10}	a^9	a^7	a^6

Note that the above two XMP-neofields don't have any stronger properties, like XIP, LXP, RXP, or CMP. They only have XMP. We now turn our attention to LXP-admissible partitions.

<u>Lemma I.26</u>: Let N_v be an LXP-neofield of order v. Then a) i), ii), iii) and b) i), ii), iii) of Lemma I.17 hold.

<u>Proof</u>: It follows from the fact that an LXP-neofield is necessarily an XMP-neofield. <u>q.e.d.</u>

Naturally, we define a pair (k,n) to be admissible (mod v-1) as we did in Definition I.18.

<u>Lemma I.27</u>: Let N_v be any LXP-neofield of even order v. Let $1+a^k = a^n$ and $1+a^s = a^t$ in N_v. Then $|\{k,n\} \cap \{s,t\}| = 0$ or 2.

<u>Proof</u>: Assume $\{k,n\} \cap \{s,t\} \neq \emptyset$. There exists $q \in \{k,n\} \cap \{s,t\}$. If q = k = s, then $n \equiv t \pmod{v-1}$. Hence $\{k,n\} = \{s,t\}$. Same argument works for the case q = n = t.

If q = k = t, then we have

$$a^s = 1+(1+a^s) = 1+a^t = 1+a^k = a^n,$$

i.e., $s \equiv n \pmod{v-1}$, hence $\{k,n\} = \{s,t\}$. This same argument works for the case q = n = s. It completes the proof. <u>q.e.d.</u>

<u>Lemma I.28</u>: Let N_v be any LXP-neofield of odd order v. Let $1+a^k = a^{n+\ell}$ and $1+a^s = a^{t+\ell}$ in N_v and $|\{k,n\}| = |\{s,t\}| = 2$. Then $|\{k,n\} \cap \{s,t\}| = 0$ or 2.

Proof: Let $q \in \{k,n\} \cap \{s,t\}$. The proof is similar to that of Lemma I.27. We only show the case when $q = k = t$. We have

$$a^s = a^{\ell}+(1+a^s) = a^{\ell}+ a^{t+\ell} = a^{\ell}(1+a^t) = a^{\ell}(1+a^k) = a^{\ell}\cdot a^{n+\ell} = a^{n+2\ell} = a^n,$$

i.e., $s \equiv n \pmod{v-1}$ again. q.e.d.

Definition I.29: For v even, a collection $L = \{\{k_1,n_1\},\ldots,\{k_{(v-2)/2},n_{(v-2)/2}\}\}$ of unordered pairs of $\mathbf{Z}_{v-1}-\{0\}$ is said to be an LXP-admissible partition of $\mathbf{Z}_{v-1}-\{0\}$ provided each pair (k_i,n_i) is admissible (mod v-1) in the sense of Definition I.18 and

i) $\bigcup_{i=1}^{(v-2)/2} \{k_i,n_i\} = \mathbf{Z}_{v-1}-\{0\}$,

ii) $\bigcup_{\{k_i,n_i\}\in L} \{\pm(n_i-k_i)\} = \mathbf{Z}_{v-1}-\{0\}$.

For v odd, a collection $L = \{\{k_1,n_1\},\{k_2,n_2\},\ldots\{k_{(v-1)/2},n_{(v-1)/2}\}\}$ of pairs of $\mathbf{Z}_{v-1}-\{\ell\}$ is said to be an LXP-admissible partition of $\mathbf{Z}_{v-1}-\{\ell\}$ provided each pair (k_i,n_i) is admissible (mod v-1) in the sense of Definition I.18, except for one pair with the same element, say $k_{(v-1)/2} = n_{(v-1)/2}$, and

i) $\left(\bigcup_{v=1}^{(v-3)/2} \{k_i,n_i\}\right) \cup \{k_{(v-1)/2}\} = \mathbf{Z}_{v-1}-\{\ell\}$,

ii) $\bigcup_{\{k_i,n_i\}\in L} \{\pm(n_i-k_i)\} = \mathbf{Z}_{v-1}-\{\ell\}$.

Lemma I.30: For v even and N_v an LXP-neofield of order v, then $\{\{k,n\} | 1+a^k = a^n\}$ is an LXP-admissible partition of $\mathbf{Z}_{v-1}-\{0\}$. For v odd and N_v an LXP-neofield of order v, then $\{\{k,n\} | 1+a^k = a^{n+\ell}\}$ is an LXP-admissible partition of $\mathbf{Z}_{v-1}-\{\ell\}$.

Proof: For v even, that each (k,n) in $\{\{k,n\} | 1+a^k = a^n\}$ is admissible follows from Lemma I.26. Since N_v has LXP, by Lemma I.16 a), $1+a^k = a^n$ implies $1+a^n = a^k$. Hence there are $\frac{v-2}{2}$ pairs in $\{\{k,n\} | 1+a^k = a^n\}$. Moreover, by Lemma I.27, all the pairs are distinct. Hence

$$\left|\bigcup_{i=1}^{(v-2)/2} \{k_i,n_i\}\right| = 2\cdot \frac{v-2}{2} = v-2,$$

i.e., $\bigcup_{i=1}^{(v-2)/2} \{k_i,n_i\} = \mathbf{Z}_{v-1}-\{0\}$. To prove ii) of Definition I.29, we will compute the cardinality of $\bigcup_{1+a^{k_i}=a^{n_i}} \{\pm(n_i-k_i)\}$. First of all, we observe that for an arbitrary pair $\{k,n\}$ in $\{\{k,n\} | 1+a^k = a^n\}$, $k-n \not\equiv n-k$. For if $k-n \equiv n-k \pmod{v-1}$, then $2k \equiv 2n \pmod{v-1}$, i.e., $2(k-n) \equiv 0 \pmod{v-1}$. Since v-1 is odd, we have $k \equiv n \pmod{v-1}$, a contradiction. Next, we observe that $\{n_1-k_1, k_1-n_1\} \cap \{n_2-k_2, k_2-n_2\} = \emptyset$ for any two pairs $\{k_1,n_1\}$, $\{k_2,n_2\}$ in $\{\{k,n\} | 1+a^k = a^n\}$. For if $n_1-k_1 = n_2-k_2$, we have $n_1+k_2 = n_2+k_1$ and

$$a^{k_2}(1+a^{k_1}) = a^{k_2}\cdot a^{n_1} = a^{k_2+n_1} = a^{n_2+k_1} = a^{n_2}\cdot a^{k_1} = a^{k_1}\cdot(1+a^{k_2}),$$

i.e., $a^{k_2}+a^{k_1+k_2} = a^{k_1}+a^{k_1+k_2}$. Hence we have $k_1 \equiv k_2 \pmod{v-1}$, a contradiction. The remaining three cases follow similarly.

The proof of the lemma for the case where v is odd is similar to the above argument. q.e.d.

The above lemma indicates that a necessary condition that an LXP-neofield of order v exists is that an LXP-admissible partition of $\mathbb{Z}_{v-1}-\{0\}$ when v is even and of $\mathbb{Z}_{v-1}-\{\ell\}$ when v is odd exists. We will show that this is also sufficient.

For an admissible pair $\{k,n\}$ in an LXP-admissible partition, we define (here $\ell = (v-1)/2$)

$$LXA(k,n) = \{(k,n),(n,k)\}, \text{ if v is even.}$$

$$LXB(k,n) = \{(k,n+\ell),(n,k+\ell)\}, \text{ if v is odd.}$$

Definition I.31: For an LXP-admissible partition L of $\mathbb{Z}_{v-1}-\{0\}$, when v is even (of $\mathbb{Z}_{v-1}-\{\ell\}$ when v is odd), let Π be defined as follows:

a) if v is even, $\Pi = \bigcup_{\{k,n\}\in L} LXA(k,n)$

b) if v is odd, $\Pi = \bigcup_{\{k,n\}\in L} LXB(k,n)$.

Theorem I.32: For $v\in\mathbb{Z}$, $v>0$, let $G = \{1,a,a^2,\ldots,a^{v-2}\}$ be the cyclic group of order v-1 and let $S = G\cup\{0\}$, with multiplication in S the extension of multiplication in G given by $0\cdot g = g\cdot 0 = 0$ for all $g\in S$. Suppose there exists an XLP-admissible partition L:

$$L = \{\{k_1,n_1\},\ldots,\{k_{(v-2)/2},n_{(v-2)/2}\}\}, \text{ if v is even,}$$

and

$$\mathcal{L} = \{\{k_1,n_1\},\ldots,\{k_{(v-1)/2},n_{(v-1)/2}\}\}, \text{ if } v \text{ is odd,}$$

and define an addition in S by (note that $\ell = (v-1)/2$)

$$\begin{cases} 0+x = x+0 = x, \quad x \in S. \\ 1+1 = 0, \text{ if } v \text{ is even; and } 1+a^{\ell} = 0, \text{ if } v \text{ is odd.} \\ 1+a^k = \begin{cases} a^n, \text{ if } (k,n) \text{ is in some LXA}(k,n) \text{ of } \pi. \\ a^{n+\ell}, \text{ if } (k,n+\ell) \text{ is in some LXB}(k,n) \text{ of } \pi. \end{cases} \\ a^r+a^s = a^r(1+a^{s-r}), \; r,s \neq 0 \end{cases}$$

Then the addition is well-defined and $N_v = \langle S,+,\cdot \rangle$ is an LXP-neofield of order v.

Proof: By Definitions I.21 and I.29, an LXP-admissible partition is also an XMP-admissible partition. Here it is clear that N_v is a cyclic neofield of order v. We only have to show the LXP-property.

For v even, $1+a^k = a^n$ implies that (k,n) is in some LXA(k,n). Hence (n,k) is also in the same LXA(k,n). Then we have $1+a^n = a^k$. Thus

$$1+(1+a^k) = 1+a^n = a^k.$$

This implies that $a^r+(A^r+a^s) = a^r(1+(1+a^{s-r})) = a^r(a^{s-r}) = a^s$, i.e., N_v has LXP property.

For v odd, $1+a^k = a^{n+\ell}$ implies that $(k,n+\ell)$ is in some LXB(k,n). Hence $(n,k+\ell)$ is also in LXB(k,n). We have $1+a^n = a^{k+\ell}$. Thus

$$a^{\ell}+(1+a^k) = a^{\ell}+a^{n+\ell} = a^{\ell}(1+a^n) = a^{\ell}\cdot a^{k+\ell} = a^k.$$

This implies that:

$$\begin{aligned}(-a^r)+(a^r+a^s) &= a^{r+\ell}+(a^r+a^s)\\ &= a^r(a^{\ell}+(1+a^{s-r}))\\ &= a^r(a^{s-r})\\ &= a^s.\end{aligned}$$

It follows that N_v has LXP property. <u>q.e.d.</u>

The general construction of LXP-neofields will be given in latter chapters. Here we give two examples to illustrate how we construct LXP-neofields from LXP-admissible partitions.

<u>Example I.33</u>: Let v = 10. We have the following LXP-admissible partition of $\mathbb{Z}_9-\{0\}$:

$$L = \{\{1,4\},\ \{2,6\},\ \{3,5\},\ \{7,8\}\}.$$

Therefore,

$$\begin{aligned}LXA(1,4) &= \{(1,4),\ (4,1)\}\\ LXA(2,6) &= \{(2,6),\ (6,2)\}\\ LXA(3,5) &= \{(3,5),\ (5,3)\}\\ LXA(7,8) &= \{(7,8),\ (8,7)\}.\end{aligned}$$

Then the presentation function is defined as:

x	0	1	a	a^2	a^3	a^4	a^5	a^6	a^7	a^8
T(x)	1	0	a^4	a^6	a^5	a	a^3	a^2	a^8	a^7

It is interesting to note that a CIP-neofield

doesn't exist for the order v = 10, (see [4]), neither does an XIP-neofield (by exhustion check). Here we have just presented an LXP-neofield of order 10.

Example I.34: Let v = 9. We have the following LXP-admissible partition of $\mathbf{Z}_8-\{\ell\}$, where $\ell = 4$,

$$L = \{\{1,2\}, \{3,6\}, \{5,7\}, \{0,0\}\}.$$

Therefore,

$$\begin{aligned} LXB(1,2) &= \{(1,6), (2,5)\} \\ LXB(3,6) &= \{(3,2), (6,7)\} \\ LXB(5,7) &= \{(5,3), (7,1)\} \\ LXB(0,0) &= \{(0,4)\}. \end{aligned}$$

Then the presentation function of this LXP-neofield of order 9 is:

x	0	1	a	a^2	a^3	a^4	a^5	a^6	a^7
$T(x)$	1	a^4	a^6	a^5	a^2	0	a^3	a^7	a .

Sec. 3. XIP-Admissible Partitions of $\mathbb{Z}_{v-1}$

We now restrict ourselves to the structure of XIP-neofields. Firstly, we have the case when v is even.

Lemma I.35: Let N_v be an XIP-neofield of even order v.

a) If g.c.d.(3,v-1) = 1, then $1+a^k = a^n$ in N_v implies that each of the conditions

i) $k \not\equiv 0 \pmod{v-1}$ and $n \not\equiv 0 \pmod{v-1}$

ii) $k \not\equiv \pm n \pmod{v-1}$

iii) $k \not\equiv 2n \pmod{v-1}$ and $n \not\equiv 2k \pmod{v-1}$

holds for the pair (k,n).

b) If $3 \mid (v-1)$, then $1+a^k = a^n$ in N_v implies that each of i), ii), and iii) in a) holds or

$$\{k,n\} = \left\{\frac{v-1}{3}, \frac{2(v-1)}{3}\right\}.$$

Proof:

a) It is obvious that $k \not\equiv 0 \pmod{v-1}$, $n \not\equiv 0 \pmod{v-1}$ and $k \not\equiv n \pmod{v-1}$.

If $k \equiv -n \pmod{v-1}$, then from a) i) and a) iii) of Lemma I.15,

$$a^n = 1+a^k = 1+a^{-n} = a^{k-n},$$

which implies that

$k-n \equiv n \pmod{v-1}$, or $3n \equiv 0 \pmod{v-1}$,

a contradiction, since g.c.d.(3,v-1) = 1. Hence $k \not\equiv -n$ (mod v-1).

If $k \equiv 2n \pmod{v-1}$, we have $n \equiv k-n \pmod{v-1}$. Then from a) i) and a) iii) of Lemma I.15, we have

$$1+a^{-n} = a^{k-n} = a^{n} = 1+a^{k}.$$

Since the addition table of N_v is a Latin square, we have

$k \equiv -n \pmod{v-1}$, or $3n \equiv 0 \pmod{v-1}$, as before.

If $n \equiv 2k \pmod{v-1}$, then from a) i) and a) ii) of Lemma I.15,

$$a^{-k} = 1+a^{n-k} = 1+a^{k} = a^{n},$$

i.e., $k \equiv -n \pmod{v-1}$, already shown to be impossible.

b) As in the proof of a) above, $k \equiv 0 \pmod{v-1}$, or $n \equiv 0 \pmod{v-1}$, or $k \equiv n \pmod{v-1}$ are not possible.

If $k \equiv -n \pmod{v-1}$, as before $3n \equiv 0 \pmod{v-1}$, so

$$n = \frac{2(v-1)}{3}, \quad k = \frac{v-1}{3}$$

or

$$k = \frac{2(v-1)}{3}, \quad n = \frac{v-1}{3}$$

are the possible solutions. Hence for $\{k,n\} \neq \left\{\frac{v-1}{3}, \frac{2(v-1)}{3}\right\}$, the conclusions of a) apply. q.e.d.

Definition I.36: For v even and (k,n) a pair of residues (mod v-1), the pair (k,n) is said to be XIP-admissible if it satisfies each of a) i), a) ii), and a) iii) of the preceding Lemma or $\{k,n\} = \left\{\frac{v-1}{3}, \frac{2(v-1)}{3}\right\}$.

Definition I.37: Denote by S(k,n) the set

$$S(k,n) = \{k,n,n-k,-k,-n,k-n\} \pmod{v-1},$$

when (k,n) is an admissible pair of residues (mod v-1).

Lemma I.38: For v even, (k,n) admissible (mod v-1) imply either $|S(k,n)| = 6$ or $\{k,n\} = \left\{\frac{v-1}{3}, \frac{2(v-1)}{3}\right\}$ and $|S(k,n)| = 2$.

Proof: (k,n) admissible and $\{k,n\} \neq \left\{\frac{v-1}{3}, \frac{2(v-1)}{3}\right\}$ implies that

$$k \not\equiv n \pmod{v-1} \text{ and } k \not\equiv -n \pmod{v-1}.$$

Also $k \not\equiv n-k \pmod{v-1}$, else $2n \equiv n \pmod{v-1}$. Clearly $k \not\equiv -k$ (mod v-1), else $2k \equiv 0 \pmod{v-1}$, i.e., $k \equiv 0 \pmod{v-1}$. And $k \not\equiv k-n$ (mod v-1), else $n \equiv 0 \pmod{v-1}$.

Similarly, apply the argument above with any element in the set S(k,n) playing the role of k. q.e.d.

Lemma I.39: Let N_v be any XIP-neofield of even order v. Let $1+a^k = a^n$ and $1+a^s = a^t$ in N_v. If $|S(k,n) \cap S(s,t)| > 2$, then $S(k,n) = S(s,t)$.

Proof: Since (k,n) and (s,t) are admissible pairs (mod v-1), we have by Lemma I.38 that $|S(k,n)| = 6$ or 2. Be hypothesis, we may assume $|S(k,n)| = |S(s,t)| = 6$.

Recall that

$$S(k,n) = \{k, n, n-k, -k, -n, k-n\}$$

and

$$S(s,t) = \{s, t, t-s, -s, -t, s-t\}.$$

Assume u, v, w,$\in S(k,n) \cap S(s,t)$. Then u, v, w$\in S(k,n)$. If none of these three elements is the negative of any of the remaining two, without loss of generality, we may assume $\{u, v, w\} = \{k, n, n-k\}$. It is obvious that $S(k,n) = S(s,t)$, since u, v, w are in $S(s,t)$ and the negatives of these three elements are also in $S(s,t)$. If one of u, v, w is the negative of the other one, say $v = -u$, then $\{u, -u, w, -w\} \subset S(k,n) \cap S(s,t)$. Hence $S(k,n) - \{u, -u, w, -w\} = \{z, -z\}$ for some $z \in S(k,n)$. But since the element z in $S(k,n)$ must be the difference of those two elements preceding z (which are distinct from $-z$), z must be the difference of two consecutive elements from the set $\{u, -u, w -w\}$ in $S(s,t)$. Therefore $\{z, -z\} \subset S(s,t)$; and then $S(k,n) = S(s,t)$. <u>q.e.d.</u>

<u>Lemma I.40</u>: For any XIP-neofield N_v of order v, let $1+a^k = a^n$ and $1+a^s = a^t$ in N_v. Then we have $|S(k,n) \cap S(s,t)| = 0, 2$ or 6.

<u>Proof</u>: If $|S(k,n) \cap S(s,t)| = 2$, then $S(k,n) = S(s,t$ by Lemma I.39. Hence $|S(k,n) \cap S(s,t)| = 6$. It remains to show that if $S(k,n) \neq S(s,t)$, then $|S(k,n) \cap S(s,t)| = 0$ or 2.

Now we assume $|S(k,n) \cap S(s,t)| \neq 0, 6$, i.e., $S(k,n)$ and $S(s,t)$ are not disjoint and are not equal. Let $x \in S(k,n) \cap S(s,t)$. Since the negative of any element mod $(v-1)$ in $S(k,n)$ is still in $S(k,n)$, $\{x,-x\} \in S(k,n) \cap S(s,t)$. Since v is even, $-x \not\equiv x \pmod{v-1}$. Thus $|S(k,n) \cap S(s,t)| = 2$. This completes the proof of the lemma. <u>q.e.d.</u>

<u>Theorem I.41</u>: If an XIP-neofield of even order v exists, then $v \equiv 2$ or $4 \pmod 6$.

<u>Proof</u>: By Lemma I.38, the collection H of all possible

S(k,n) corresponding to additions $1+a^k = a^n$ in N_v consists of sets all of which have cardinality 6 or all but one of which has cardinality 6, that exception having cardinality 2 If these sets partition $\mathbf{Z}_{v-1}-\{0\}$, then clearly $v \not\equiv 2$ or 4 (mod 6). If some of the S(k,n)'s are not disjoint, then we shall discuss the following:

Since the sextuples reduce to cardinality 2 only when $v \equiv 4 \pmod 6$, we may assume $|S(k,n)| = 6$ for all admissible pair (k,n). Moreover, we assume there are t totally disjoint sextuples ($t \neq 0$). (Call this subcollection $H_t \subset H$, i.e., any two sextuples in H_t are disjoint and each sextuple in H_t is disjoint from all sextuples in $H-H_t$.)

Let x be the number of elements in $\mathbf{Z}_{v-1}-\{0\}$ which appear only in one sextuple of H. Let y be the number of elements in $\mathbf{Z}_{v-1}-\{0\}$ which appear in exactly two sextuples of the collection. Since the addition table of N_v is a Latin square, no element of $\mathbf{Z}_{v-1}-\{0\}$ appear in more than two sextuples of H. Now if $v \equiv 0 \pmod 6$, we have $x+y = 4 \pmod 6$ by counting all the elements in $\mathbf{Z}_{v-1}-\{0\}$. But on the other hand, $|\bigcup_i S(k_i,n_i)| = x+2y \equiv 0 \pmod 6$ by counting the multiplicities. Solving this system of congruence equations, we have $x \equiv 2 \pmod 6$. This is a contradiction since $x = 6t$. <u>q.e.d.</u>

Remark that if $t = 0$, then $x = 0$ and $2y \equiv 0 \pmod 6$. Since $y = v-2$, it must be even. Hence $y \equiv 0 \pmod 3$ and $y \equiv 0 \pmod 2$, <u>i.e.</u>, $y \equiv 0 \pmod 6$. Then $v = y+2 \equiv 2 \pmod 6$.

<u>Definition I.42</u>: For (k,n) an admissible pair of residues (mod v-1) and S(k,n) the set

$$S(k,n) = \{k,\ n,\ n-k,\ -k,\ -n,\ k-n\} \pmod{v-1},$$

with $|S(k,n)| = 6$, denote by <u>OS(k,n)</u> the set

$$OS(k,n) = \{k,\ n-k,\ -n\},$$

and by <u>ES(k,n)</u> the set

$$ES(k,n) = \{n,\ -k,\ k-n\}.$$

They are called the (set of) <u>odd-parity elements</u> and <u>even-parity elements</u> respectively of the set S(k,n).

<u>Definition I.43</u>: Call a collection $\{S(k_1,n_1),\ S(k_2,n_2),\ \ldots,\ S(k_h,n_h)\} = \{S_1,\ S_2,\ \ldots,\ S_h\}$ of sets $S(k_i,n_i)$ defined in Definition I.37 an <u>XIP-admissible partition of $\mathbf{Z}_{v-1}^*$</u> provided each pair $(k_1,n_1),\ \ldots,\ (k_h,n_h)$ is XIP-admissible (mod v-1) in the sense of Definition I.36 and either

a) $\{S(k_1,n_1),\ \ldots,\ S(k_h,n_h)\}$ is a partition of $\mathbf{Z}_{v-1}^*$,

or

b) $\bigcup_{i=1}^{h} S(K_i,n_i) = \mathbf{Z}_{v-1}^*$ and $\{S_1,\ \ldots,\ S_h\} = H_t \cup \{S_{t+1},\ \ldots,\ S_h\}$

where i) H_t is the subcollection of t totally disjoint sextuples (<u>cf</u>: Theorem I.42, proof)

ii) Each element d of $\mathbf{Z}_{v-1}^* - \bigcup_{i=1}^{t} S_i$ appears in exactly two sextuples of the subcollection $H-H_t$ and d is of odd-parity in one sextuple and is of even-parity in the other.

iii) Each element of $\mathbb{Z}_{v-1}{}^* - \bigcup_{i=t+1}^{h} S_i$ appears in exactly one sextuple in H_t.

Lemma I.44: For v even and N_v an XIP-neofield of order v, then $\{S(k,n) \mid 1+a^k = a^n\}$ is an XIP-admissible partition of $\mathbb{Z}_{v-1}-\{0\}$ of either type a) or type b) defined in the preceding definition.

Proof: Recall that in the proof of Theorem I.42, if those sextuples in H partition $\mathbb{Z}_{v-1}-\{0\}$, then we have an XIP-admissible partition of type a). If they don't partition $\mathbb{Z}_{v-1}-\{0\}$, then we have an XIP-admissible partition of type b). q.e.d.

The above lemma indicates that a necessary condition that an XIP-neofield of even order v exists is that an XIP-admissible partition of $\mathbb{Z}_{v-1}-\{0\}$ of type a) or type b) exists. We will show that either of these conditions is also sufficient.

For $S(k,n) = \{k, n, n-k, -k, -n, k-n\}$ where (k,n) is admissible, denote by CIA(k,n) and SIA(k,n) the collections:

$$CIA(k,n) = \{(k,n), (n-k,-k), (-n,k-n), (n,k), (-k,n-k), (k-n,-n)\}$$

and

$$SIA(k,n) = \{(k,n), (n-k,-k), (-n,k-n), (n,n-k), (-k,-n), (k-n,k)\}$$

of ordered pairs of residues (mod v-1).

<u>Definition I.45</u>: For an XIP-admissible partition of $\mathbb{Z}_{v-1}-\{0\}$ of type a) or type b), $\{S(k,n)\}$, let $\pi = \{XIA(k,n)\}$ be defined as either of the following four cases:

a) If $\{S(k,n)\}$ is of type a), then either

i) all XIA(k,n)'s are of type CIA(k,n),

or

ii) all XIA(k,n)'s are of type SIA(k,n),

or

iii) for each admissible pair $(k,n) \in \mathbb{Z}_{v-1}{}^* \times \mathbb{Z}_{v-1}{}^*$ where $\mathbb{Z}_{v-1}{}^* = \mathbb{Z}_{v-1}-\{0\}$, in the XIP-admissible partition $\{S(k,n)\}$, either XIA(k,n) = CIA(k,n) or XIA(k,n) = SIA(k,n), but excluding the cases i) and ii) above.

b) If $\{S(k,n)\}$ is of type b), then $XIA(k,n) = \{(k,n), (n-k,-k), (-n,k-n)\}$ for all (k,n) when t = 0. When $t \neq 0$, $XIA(k,n) = \{(k,n), n-k,-k), (-n,k-n)\}$ for those (k,n) such that $S(k,n) \in \{S_{t+1}, \ldots, S_h\}$ and XIA(k,n) = CIA(k,n) or SIA(k,n) for those (k,n) such that $S(k,n) \in H_t$.

<u>Theorem I.46</u>: For $v \equiv 2$ or 4 (mod 6), $v \in \mathbb{Z}$, $v > 0$, let $G = \{1, a, a^2, a^3, \ldots, a^{v-2}\}$ be the cyclic group of order v-1, and let $S = G \cup \{0\}$, with multiplication in S the extension of multiplication in G given by $0 \cdot g = g \cdot 0 = 0$ for all $g \in S$. Suppose there exists an XIP-admissible partition π of either type a) or type b),

$S(k_1 n_1), \ldots, S(k_n, n_h)$ of $\mathbb{Z}_{v-1} - \{0\}$,

and define an addition in S by

$$\begin{cases} 0+x = x+0 = x, \quad x \in S. \\ 1+1 = 0 \\ 1+a^k = a^n, \text{ provided } (k,n) \text{ is in some} \\ \qquad\qquad XIA(k,n) \text{ of } \pi \\ a^r+a^s = a^r(1+a^{s-r}), \; r,\, s \neq 0. \end{cases}$$

Then this addition is well-defined and $N_v = \langle S,+,\cdot\rangle$ is an XIP-neofield of order v.

Proof: For the XIP-admissible partition of type a) i), it is proved in [4] (Theorem I.9) that $\langle S,+,\cdot\rangle$ is a CIP-neofield. Hence it is also an XIP-neofield.

We prove the case a) ii) as follows. Since π is a partition of $\mathbb{Z}_{v-1} - \{0\}$, for each $k \in \mathbb{Z}_{v-1} - \{0\}$, there is a unique pair in some $XIA(k_i, n_i)$ for $i \in \{1, 2, 3, 4, \ldots, h\}$ and thus $1+a^k$ is defined for all $a^k \in G$. Hence the addition table for every pair of $S \times S$ is well-defined.

Label the rows and columns of the $(v \times v)$- addition table M of $\langle S,+\rangle$ in such a way that the (x,y) entry is the sum x+y.

From the defining relation, the 0-row of M contains all elements of S. The 1-row contains 1 in the 0-column and 0 in the 1-column. Each $n \in \mathbb{Z}_{v-1} - \{0\}$ occurs exactly once as a second element in some ordered pair (k,n) of $XIA(k_i, n_i)$, so the remaining elements of the 1-row consist of $a, a^2, \ldots, a^{v-2}$ in some order, i.e., the 1-row contains each element of S exactly once.

By the defining relation again, the elements of the a^k-th row of M can be obtained from the elements of the 1-row by multiplying the elements in the 1-row by a^k; hence all elements of S occur in each row of M. Similarly, all the elements in each column are distinct, hence $\langle S,+\rangle$ is a loop.

Distributivity of multiplication over addition follows from $a^r+a^s = a^r(1+a^{s-r})$ and the commutativity of multiplication.

Now we will show that $\langle S,+,\cdot\rangle$ has XIP. If $1+a^k = a^n$ in $\langle S,+,\cdot\rangle$, by defining relation, we have a positive integer i such that $(k,n)\in XIA(k_i,n_i)$ and $S(k,n) = S(k_i,n_i)$. It then follows that

$$a^k+(1+a^k) = a^k+a^n = a^k(1+a^{n-k}) = a^k\cdot a^{-k} = 1,$$

by using $(n-k,-k)\in XIP(k_i,n_i)$. Applying distributive law, we have $a^ra^k+(a^r+a^ra^k) = a^r$. Hence if $s = r+k$, we have $a^s+(a^r+a^s) = a^r$. Since r,s are arbitrary in $\mathbf{Z}_{v-1}-\{0\}$, we have proved that $\langle S,+,\cdot\rangle$ has XIP.

We prove case a) iii): The proof that $\langle S,+\rangle$ is a loop is similar to what we did above. Now we show that $N_v = \langle S,+,\cdot\rangle$ has XIP. It suffices to show that $(-x)+(1+x) = 1$, $\forall x\in N_v$. If $1+a^k = a^n$, then $(k,n)\in XIA(k_i,n_i)$ for some i. Since $S(k,n) = S(k_i,n_i)$, it follows that $XIA(k_i,n_i) = XIA(k,n)$, so $(n-k,-k)\in XIA(k_i,n_i)$. Then

$$(-a^k)+(1+a^k) = a^k+a^n = a^k(1+a^{n-k}) = a^k \cdot a^{-k} = 1$$

in N_v. So $N_v = \langle S,+,\cdot \rangle$ has XIP.

Finally we prove case b). First of all, we assume $t = 0$. By Definition I.42, each element d of $\mathbb{Z}_{v-1} - \{0\}$ appears in exactly two sextuples in the collection $\{S(k_1,n_1), \ldots, S(k_h,n_h)\}$, where the element is of odd-parity in one sextuple and of even-parity in the second sextuple. This would determine the second row of the addition table in such a way that all elements of the second row are distinct. By Definition I.42 again, $\bigcup_{(k,n)} OS(k,n) = \mathbb{Z}_{v-1} - \{0\}$. We have also that

$$\bigcup_{XIA(k,n)} \{ y-x \mid (x,y) \in XIA(k,n)\} = \bigcup_{(k,n)} OS(k,n)$$
$$= \mathbb{Z}_{v-1} - \{0\}.$$

It follows that $|\{n-k \mid 1+a^k = a^n\}| = v-2$. Hence the addition table (after applying distributive law) is a Latin square. That $N_v = \langle S,+,\cdot \rangle$ has XIP follows from the defining relation that

$$XIA(k,n) = \{(k,n), (n-k,-k), (-n,k-n)\},$$

which is of XIP property.

Next, we assume $t \neq 0$. By Definition I.42, the number of exponents of the second row of the addition table is

$$6t+3 \cdot \frac{(v-2-6t)x2}{6} = 6t+(v-2-6t) = v-2,$$

where $v-2-6t$ elements of $\mathbb{Z}_{v-1} - \{0\}$ appear in $\bigcup_{i=t+1}^{h} S_i$

with multiplicities 2 (and then we have $\frac{(v-2-6t)\times 2}{6}$ sextuples of S_i not in H_t, with each of the sextuples contributing three exponents). Since these v-2 elements are distinct (mod v-1), together with 1+0 = 1 and 1+1 = 0, we have that all elements in the second row of the addition table are distinct. Moreover, $\left|\bigcup_{i=t+1}^{h} OS(k_i,n_i)\right| =$ (v-2)-6t and

$$\begin{aligned}
&\#\left(\bigcup_{XIA(k,n)}\{y-x \mid (x,y)\in XIA(k,n)\}\right)\\
&= \#\left(\bigcup_{i=t+1}^{h}\{y-x \mid (x,y)\in XIA(k_i,n_i)\}\right)+\#\left(\bigcup_{i=1}^{t}\{y-x \mid (x,y)\in XIA(k_i,n_i)\}\right)\\
&= \#\left(\bigcup_{i=t+1}^{h} OS(k_i,n_i)\right)+\#\left(\bigcup\{y-x \mid S(x,y)\in H_t\}\right)\\
&= [(v-2)-6t]+6t\\
&= v-2.
\end{aligned}$$

Hence, after applying the distributive law, the addition table is a Latin square. Thus $\langle S,+,\cdot\rangle$ is a cyclic neofield. To show that $\langle S,+,\cdot\rangle$ has XIP, we simply make the following observation. The addition table defined from those S(k,n) in H_t has either CIP or just XIP. The defining relation from those S(k,n) not in H_t has just XIP. <u>q.e.d.</u>

By the above theorem, we have three types of proper XIP's of even order (which are not CIP-neofields), one with type a) ii), one with type a) iii) and the last one with type b Those proper XIP-neofields of types a) ii) and a) iii) will be discussed in Chapter II and examples will be given there. The construction of proper XIP-neofields of type b) will be given in Chapter III. Here we give three examples to illustrate proper XIP-neofields of even order of type b), since type b) is more interesting.

<u>Example I.47</u>: Let v = 14. We have the following XIP-admissible partition of type b) of $\mathbb{Z}_{13}-\{0\}$.

$$(1,4) = \{1, 4, 3, 12, 9, 10\}$$
$$(2,6) = \{2, 6, 4, 11, 7, 9\}$$
$$(5,2) = \{5, 2, 10, 8, 11, 3\}$$
$$(6,1) = \{6, 1, 8, 7, 12, 5\}.$$

Therefore the additive relations are given by the following four sets of ordered pairs:

$$XIA(1,4) = \{(1,4), (3,12), (1,10)\}$$
$$XIA(2,6) = \{(2,6), (4,11), (7,9)\}$$
$$XIA(5,2) = \{(5,2), (10,8), (11,3)\}$$
$$XIA(6,1) = \{(6,1), (8,7), (12,5)\}.$$

Then the presentation function of this proper XIP-neofield N_{14} of type b) is:

x	0	1	a	a^2	a^3	a^4	a^5	a^6	a^7	a^8	a^9	a^{10}	a^{11}	a^{12}
$T(x)$	1	0	a^4	a^6	a^{12}	a^{11}	a^2	a	a^9	a^7	a^{10}	a^8	a^3	a^5

.

<u>Example I.48</u>: Let v = 16. We have the following XIP-admissible partition of $\mathbb{Z}_{15}-\{0\}$ of type b):

$$(1,7) = \{1,\ 7,\ 6,\ 14,\ 8,\ 9\}$$
$$(7,3) = \{7,\ 3,\ 11,\ 8,\ 12,\ 4\}$$
$$(3,1) = \{3,\ 1,\ 13,\ 12,\ 14,\ 2\}$$
$$(9,11) = \{9,\ 11,\ 2,\ 6,\ 4,\ 13\}$$
$$(5,10) = \{5,10\}.$$

The additive relations are given by:

$$XIA(1,7) = \{(1,7),\ (6,14),\ (8,9)\}$$
$$XIA(7,3) = \{(7,3),\ (11,8),\ (12,4)\}$$
$$XIA(3,1) = \{(3,1),\ (13,12),\ (14,2)\}$$
$$XIA(9,11) = \{(9,11),\ (2,6),\ (4,13)\}$$
$$XIA(5,10) = \{(5,10),\ (10,5)\}.$$

Hence we have the presentation function:

x	0	1	a	a^2	a^3	a^4	a^5	a^6	a^7	a^8	a^9	a^{10}	a^{11}	a^{12}	a^{13}	a^{14}
T(x)	1	0	a^7	a^6	a	a^{13}	a^{10}	a^{14}	a^3	a^9	a^{11}	a^5	a^8	a^4	a^{12}	a^2 .

<u>Example I.49</u>: Let $v = 20$. We have the following XIP-admissible partition of $\mathbb{Z}_{19}-\{0\}$ of type b):

$$(5,2) = \{5,\ 2,\ 16,\ 14,\ 17,\ 3\}$$
$$(2,8) = \{2,\ 8,\ 6,\ 17,\ 11,\ 13\}$$
$$(8,1) = \{8,\ 1,\ 12,\ 11,\ 18,\ 7\}$$
$$(1,5) = \{1,\ 5,\ 4,\ 18,\ 14,\ 15\}$$
$$(3,12) = \{3,\ 12,\ 9,\ 16,\ 7,\ 10\}$$
$$(10,4) = \{10,\ 4,\ 13,\ 9,\ 15,\ 6\}.$$

The additive relations are given by:

$$XIA(5,2) = \{(5,2),\ (16,14),\ (17,3)\}$$

$$XIA(2,8) = \{(2,8),\ (6,17),\ (11,13)\}$$

$$XIA(8,1) = \{(8,1),\ (12,11),\ (18,7)\}$$

$$XIA(1,5) = \{(1,5),\ (4,18),\ (14,15)\}$$

$$XIA(3,12) = \{(3,12),\ (9,16),\ (7,10)\}$$

$$XIA(10,4) = \{(10,4),\ (13,9),\ (15,6)\}.$$

Hence the presentation function is:

x	0	1	a	a^2	a^3	a^4	a^5	a^6	a^7	a^8	a^9	a^{10}	a^{11}	a^{12}	a^{13}
$T(x)$	1	0	a^5	a^8	a^{12}	a^{18}	a^2	a^{17}	a^{10}	a	a^{16}	a^4	a^{13}	a^{11}	a^9

x	a^{14}	a^{15}	a^{16}	a^{17}	a^{18}
$T(x)$	a^{15}	a^6	a^{14}	a^3	a^7 .

We now turn our attention to the case of odd order XIP-neofields.

<u>Definition I.50</u>: For v odd, let $\ell = \frac{v-1}{2}$ and let

$$S(k,n) = \{k,\ n,\ n-k,\ -k,\ -n,\ k-n\},$$

$$CIB(k,n) = \{(k,n+\ell),\ (n-k,-k+\ell),\ (-n,k-n+\ell),\ (n,k+\ell),\ (-k,n-k+\ell),\ (k-n,-n+\ell)\}$$

$$SIB(k,n) = \{(k,n+\ell),\ (n-k,-k+\ell),\ (-n,k-n+\ell),\ (n,n-k+\ell),\ (-k,-n+\ell),\ (k-n,k+\ell)\}$$

$$XIB(k,n) = \{(k,n+\ell),\ (n-k,-k+\ell),\ (-n,k-n+\ell)\}.$$

<u>Lemma I.51</u>: Let N_v be an XIP-neofield of odd order v, with multiplicative group $\{1,\ a,\ a^2.\ a^3,\ \ldots\ a^{v-2}\}$ and suppose $k,n \in \mathbb{Z}_{v-1}$, such that $1+a^k = a^{n+\ell}$ in N_v. Then $n \not\equiv \ell$ (mod $v-1$), $k \not\equiv \ell$ (mod $v-1$) and $k \not\equiv n+\ell$ (mod $v-1$).

If in addition, $n \not\equiv 0 \pmod{v-1}$, $k \not\equiv 0 \pmod{v-1}$, and $k \not\equiv \pm n \pmod{v-1}$, then $|S(k,n)| = 6$.

<u>Proof</u>: That $n \not\equiv \ell$, $k \not\equiv \ell$ and $k \not\equiv n+\ell \pmod{v-1}$ are obvious. Now assume the other conditions. Hence in $S(k,n) = \{k, n, n-k, -k, -n, k-n\}$, we have $k \not\equiv n \pmod{v-1}$, $k \not\equiv -k \pmod{v-1}$ and $k \not\equiv -n \pmod{v-1}$. If $k \equiv n-k \pmod{v-1}$, then $2k \equiv n \pmod{v-1}$, so from b) of Lemma I.15,

$$a^{-k+\ell} = 1+a^{n-k} = 1+a^{k} = a^{n+\ell},$$

implies that $k \equiv -n \pmod{v-1}$, a contradiction.

If $k \equiv k-n \pmod{v-1}$, then $n \equiv 0 \pmod{v-1}$, a contradiction. Hence k is distinct from the remaining elements of $S(k,n)$.

By using analogous method, we check all the remaining cases and conclude that $|S(k,n)| = 6$. <u>q.e.d.</u>

Given v odd and $1+a^k = a^{n+\ell}$ in the XIP-neofield N_v, we can relax any of the conditions mentioned in the above lemma, i.e., $n \not\equiv 0$, $k \not\equiv 0$, $k \not\equiv \pm n \pmod{v-1}$.

<u>Case 1</u>: $k \equiv 0 \pmod{v-1}$.

Then

$$CIB(0,n) = \{(0,n+\ell), (-n,-n+\ell), (n,\ell)\}$$

$$SIB(0,n) = \{(0,n+\ell), (-n,-n+\ell), (n,\ell), (0,-n+\ell), (-n,\ell), (n,n+\ell)\}$$

and $$XIB(0,n) = \{(0,n+\ell), (n,\ell), (-n,-n+\ell)\}.$$

It is observed that, if $n \not\equiv 0 \pmod{v-1}$, then since $n \not\equiv \ell$ $\pmod{v-1}$ the three pairs of CIB(0,n) are distinct. If $n \equiv 0 \pmod{v-1}$, then CIB(0,n) reduces to the singleton $\{(0,\ell)\}$. For XIB(0,n), we have the same situation as CIB(0,n). The SIB(0,n) is reduced to the singleton $SIB(0,0) = \{(0,\ell)\}$ when $n \equiv 0$ or $k \equiv 0$, or $k \equiv n \pmod{v-1}$, since (n,ℓ) and $(n,n+\ell)$ are excluded by one another. For XIB(k,n) and CIB(k,n) when $k \equiv n \pmod{v-1}$, it is the same as the case $k \equiv 0 \pmod{v-1}$.

<u>Case 2</u>: $k \equiv -n \pmod{v-1}$.

From the additive relations i) and iii) of Lemma I.15 b),

$$a^{n+\ell} = 1+a^{k} = 1+a^{-n} = a^{k-n+\ell}$$

So $n \equiv k-n \equiv -2n \pmod{v-1}$, i.e., $3n \equiv 0 \pmod{v-1}$. If g.c.d.$(3,v-1) = 1$, this reduces to Case 1. So $3 \mid (v-1)$ and

$$n = \frac{v-1}{3},\ k = \frac{2(v-1)}{3}$$

or

$$n = \frac{2(v-1)}{3},\ k = \frac{v-1}{3}$$

Then $XIB(k,n) = \left\{\left(\frac{v-1}{3}, \frac{v-1}{6}\right), \left(\frac{2(v-1)}{3}, \frac{5(v-1)}{6}\right)\right\}$. The cases for CIB(k,n) and SIB(k,n) are the same.

<u>Theorem I.52</u>: Let N_v be an XIP-neofield of odd order v, with $\ell = \frac{v-1}{2}$:

if $v \equiv 3 \pmod 6$, the sets XIB(k,n) defined above, determined by the addition $1+a^k = a^{n+\ell}$ in N_v are of cardinality 3, except for the singleton $XIB(0,0) = \{(0,\ell)\}$;

if $v \equiv 5 \pmod 6$, the sets XIB(k,n), determined by the addition $1+a^k = a^{n+\ell}$ in N_v, are all of cardinality 3 with one special triple

$$XIB(k,n) = \{(0,n+\ell),\ (n,\ell),\ (-n,-n+\ell)\},\ n \not\equiv 0,\ell;$$

if $v \equiv 1 \pmod 6$, the sets XIB(k,n), determined by $1+a^k = a^{n+\ell}$ in N_v, are all of cardinality 3 with possibly one special triple

$$XIB(k,n) = \{(0,n+\ell),\ (n,\ell),\ (-n,-n+\ell)\},\ n \not\equiv 0,\ell,$$

except for a pair:

$$XIB\left(\frac{v-1}{3}, \frac{2(v-1)}{3}\right) = \left\{\left(\frac{v-1}{3}, \frac{v-1}{6}\right), \left(\frac{2(v-1)}{3}, \frac{5(v-1)}{6}\right)\right\}.$$

<u>Proof</u>: If $1+a^k = a^{n+\ell}$ in N_v, then the pairs in XIB(k,n) are the exponent pairs associated with the three additions equivalent to $1+a^k = a^{n+\ell}$ by Lemma I.15, b). We will force the conclusion of this theorem by counting the exponents $n+\ell$ of elements $a^{n+\ell}$ of N_v which occur in the sums

$$1+a^k = a^{n+\ell} \quad \text{in } N_v;$$

those exponents occur as right hand entries of the ordered pairs in the sets in $D = \{XIB(k,n) \mid 1+a^k = a^{n+\ell}\}$ in N_v. Two elements never occur in the role of $a^{n+\ell}$, i.e., 1 and 0, so when $v \equiv i \pmod 6$, we require that

$$\sum_{XIB(k,n)\in D} |XIB(k,n)| \equiv i-2 \pmod 3.$$

Furthermore, the possibilities for $|XIB(k,n)| < 3$ are limited in D to at most one pair, and at most one singleton, and the existence of the singleton excludes both the special triple and the pair from arising.

Hence for $v \equiv 3 \pmod 6$,

$$\sum_{XIB(k,n)\in D} |XIB(k,n)| = v-2 \pmod 3,$$

which in light of the last paragraph, can arise only if all $XIB(k,n)\in D$ have cardinality 3, except for the singleton $XIB(0,0) = \{(0,\ell)\}$.

For $v \equiv 5 \pmod 6$, we have

$$\sum_{XIB(k,n)\in D} |XIB(k,n)| = v-2 \equiv 0 \pmod 3;$$

since the pair $XIB(\frac{v-1}{3}, \frac{2(v-1)}{3})$ and the singleton XIB(0,0) can't occur together, we conclude that all XIB(k,n) in D have cardinality 3 with one special triple

$$XIB(0,n) = \{(0,n+\ell), (-n,-n+\ell), (n,\ell)\}, \ n \notin \{0,\ell\}.$$

For $v \equiv 0 \pmod 6$, similarly we have

$$\sum_{XIB(k,n)\in D} |XIB(k,n)| = v-2 \equiv 2 \pmod 3,$$

which requires that D contain a pair

$$XIB\left(\frac{v-1}{3}, \frac{2(v-1)}{3}\right) = \left\{\left(\frac{v-1}{3}, \frac{v-1}{6}\right), \left(\frac{2(v-1)}{3}, \frac{5(v-1)}{6}\right)\right\}$$

and possibly one special triple

$$XIB(k,n) = \{(0,n+\ell), (n,\ell), (-n,-n+\ell)\}, \ n \not\equiv 0,\ell.$$

q.e.d.

Definition I.53: For v odd, a pair (k,n) of residues (mod v-1) is XIP- admissible provided that for $\ell = \frac{v-1}{2}$, $n \not\equiv \ell$ (mod v-1), $k \not\equiv \ell$ (mod v-1), $k \not\equiv n+\ell$ (mod v-1) and either

$$\text{i)} \quad \begin{cases} n \not\equiv 0 \pmod{v-1} \\ k \not\equiv 0 \pmod{v-1} \\ k \not\equiv \pm n \pmod{v-1}, \end{cases}$$

or

$$\text{ii)} \quad \begin{cases} k \equiv 0 \pmod{v-1} \\ n \not\equiv 0 \pmod{v-1}, \end{cases}$$

or

$$\text{iii)} \quad \begin{cases} k \not\equiv 0 \pmod{v-1} \\ n \equiv 0 \pmod{v-1}, \end{cases}$$

or

$$\text{iv)} \quad \begin{cases} k \equiv n \pmod{v-1} \\ n \not\equiv 0 \pmod{v-1}, \end{cases}$$

or

$$\text{v)} \quad \begin{cases} k \equiv \frac{v-1}{3} \pmod{v-1} \\ n \equiv \frac{2(v-1)}{3} \pmod{v-1}. \end{cases}$$

Definition I.54: For (k,n) an admissible pair of residues (mod v-1) and $|S(k,n)| = 6$, denote by OS(k,n) the set

$$OS(k,n) = \{k, n-k, -n\}$$

and by ES(k,n) the set

$$ES(k,n) = \{n, -k, k-n\}.$$

They are called the (set of) odd-parity elements and even-parity elements respectively of the set S(k,n).

Definition I.55: Call a collection $\{S(k_1,n_1), \ldots, S(k_h,n_h)\} = \{(S_1, \ldots, S_h)\}$, with (k_i, n_i) XIP-admissible pair defined in Definition I.53, an XIP-admissible partition of $\mathbb{Z}_{v-1}{}^*$ provided either

a) $\{S(k_1,n_1), \ldots, S(k_h,n_h)\}$ is a partition of $\mathbb{Z}_{v-1}{}^*$,

or

b) $\bigcup_{i=1}^{h} S(k_i,n_i) = \mathbb{Z}_{v-1}{}^*$ and $\{S_1, \ldots, S_{h'}, \|S_i| = 6\} = H_t \cup \{S_{t+1}, \ldots, S_{h'}\}$, $t \leq h' < h$,

where

i) H_t is the subcollection of t totally disjoint sextuples defined before in the even order case.

ii) each element d in $\mathbb{Z}_{v-1}{}^* - \bigcup_{i=1}^{t} S_i - \bigcup_{i=h'+1}^{h} S_i$ appears exactly two sextuples in the collection $\{S_{t+1}, S_{t+2}, \ldots, S_{h'}\}$ and d is of odd-parity in one sextuple and of even-parity in the other.

iii) sextuples in H_t partition $\mathbb{Z}_{v-1}{}^* - \bigcup_{i=t+1}^{h} S_i$.

Lemma I.56: Let N_v be an XIP-neofield of odd order v. Then $1+a^k = a^{n+\ell}$ and $1+a^s = a^{t+\ell}$ in N_v with $|S(k,n)| = |S(s,t)| = 6$ implies that $|S(k,n) \cap S(s,t)| = 0, 2,$ or 6.

Proof: The proof is similar to Lemma I.40 in the case when v is even. q.e.d.

Lemma I.57: For v odd and N_v an XIP-neofield of order v, $\{S(k,n) \mid 1+a^k = a^{n+\ell}\}$ is an XIP-admissible partition of

$\mathbb{Z}_{v-1}-\{\ell\}$ of either type a) or type b) defined in Lemma I.55.

The proof of the above lemma is similar to Lemma I.44 in the case when v is even.

Definition I.58: For an XIP-admissible partition of $\mathbb{Z}_{v-1}-\{\ell\}$ of type a) or type b), $\{S(k,n)\}$, let XIB = $\{XIB(k,n)\}$ be defined as one of the following four cases:

a) if $\{S(k,n)\}$ is of type a), then either

i) all the XIB(k,n)'s are of type CIB(k,n)

or

ii) all the XIB(k,n)'s are of type SIB(k,n)

or

iii) XIB(k,n) = CIB(k,n) or SIB(k,n) but excluding i) and ii) above.

b) if $\{S(k,n)\}$ is of type b), then XIB(k,n) = $\{(k,n+\ell), (n-k,-k+\ell), (-n,k-n+\ell)\}$ for all (k,n) admissible (mod v-1), when t = 0. When $t \neq 0$, then XIB(k,n) = $\{(k,n+\ell), (n-k,-k+\ell), (-n,k-n+\ell)\}$ for those (k,n) such that

$S(k,n) \in \{S_{t+1}, \ldots, S_h\} - \{S_i \mid |S_i| = 2\}$(If $|S_i| = 2$ in I.5 we have XIB(k,n) = $\{(k, n+\ell), (n, k+\ell)\}$), and XIB(k,n) = CIB(k,n) or SIB(k,n) for those (k,n) such that $S(k,n) \in H_t$.

We are ready for the following theorem.

Theorem I.59: For v odd, let $\ell = \frac{v-1}{2}$ and G = $\{1, a, a^2, \ldots, a^{v-2}\}$ be the cyclic group of order v-1 and let

$S = G\cup\{0\}$ with multiplication in S defined by

$$0\cdot g = g\cdot 0 = 0, \forall g\in S.$$

Suppose there exists an XIP-admissible partition of $\mathbb{Z}_{v-1}-\{\ell\}$, say $S(k_1,n_1), \ldots, S(k_q,n_q)$. Define an addition on S by

$$\begin{cases} 0+x = x+0 = x \\ 1+a^{(v-1)/2} = 0 \\ 1+a^k = a^n \text{ for } (k,n)\in XIB(k_i,n_i) \\ a^r+a^s = a^r(1+a^{s-r}) \text{ for } r>0. \end{cases}$$

Then $\langle S,+,\cdot\rangle$ is well-defined and $N_v = \langle S,+,\cdot\rangle$ is an XIP-neofield.

Proof: The proof is analogous to the proof of Theorem I.46 except for verifying that addition has XIP property.

To verify the XIP property, note that if $(k,n+\ell)\in XIB(k_i,n_i)$ for some $i\in\{1, 2, \ldots, q\}$, then also $(n-k,-k+\ell)\in XIB(k_i,n_i)$ and $(-n,k-n+\ell)\in XIB(k_i,n_i)$, so

$$\begin{aligned}(-a^k)+(1+a^k) &= a^{k+\ell}+a^{n+\ell} \\ &= a^{k+\ell}[1+a^{n-k}] \\ &= a^{k+\ell}(a^{-k+\ell}) \\ &= 1\end{aligned}$$

In general, we have

$$\begin{aligned}(-a^k)+(a^n+a^k) &= (-a^k)+a^n(1+a^{k-n}) \\ &= a^n[(-a^{k-n})+(1+a^{k-n})] = a^n.\end{aligned}$$

Hence $\langle S,+,\cdot\rangle$ is an XIP-neofield. q.e.d.

By Lemma I.11, Lemma I.14, Definition I.55, Definition I.58 and the above theorem, an XIP-admissible partition of Z^*_{v-1}, v odd, of type a)i) defines a CIP-neofield of odd order v. It will be clear later that we have three types of proper XIP-neofield of odd order (which are not CIP), one is type a) ii), one is type a) iii), and one is type b). Those proper XIP of type a) ii) and a) iii) will be discussed in Chapter II and those of type b) in Chapter III. Here we give three examples to illustrate proper XIP-neofields.

Example I. 60: Let v = 15, then $Z_{14}-\{\ell\} = \{0, 1, 2, 3, \cdots, 12, 13\}$, where $\ell = 7$. The sextuples are given by

$$(1, 4) = \{1, 4, 3, 13, 10, 11\}$$
$$(2, 6) = \{2, 6, 4, 12, 8, 10\}$$
$$(5, 2) = \{5, 2, 11, 9, 12, 3\}$$
$$(6, 1) = \{6, 1, 9, 8, 13, 5\}$$
$$(0, 0) = \{0\}.$$

Hence we have

$$XIB(1, 4) = \{(1, 11), (3, 6), (10, 4)\}$$
$$XIB(2, 6) = \{(2, 13), (4, 5), (8, 3)\}$$
$$XIB(5, 2) = \{(5, 9), (11, 2), (12, 10)\}$$
$$XIB(6, 1) = \{(6, 8), (9, 1), (13, 12)\}$$
$$XIB(0, 0) = \{(0, 7)\}$$

Hence we have the presentation function of the proper XIP-neofield of type b).

x	0	1	a	a^2	a^3	a^4	a^5	a^6	a^7	a^8	a^9	a^{10}	a^{11}	a^{12}	a^{13}
$T(x)$	1	a^7	a^{11}	a^{13}	a^6	a^5	a^9	a^8	0	a^3	a	a^4	a^2	a^{10}	a^{12}

Example I.61: Let $v = 21$, then $\ell = 10$. We have the following sextuples

$$(1,4) = \{1, 4, 3, 19, 16, 17\}$$
$$(2,8) = \{2, 8, 6, 18, 12, 14\}$$
$$(4,11) = \{4, 11, 7, 16, 9, 13\}$$
$$(5,2) = \{5, 2, 17, 15, 18, 3\}$$
$$(8,1) = \{8, 1, 13, 12, 19, 7\}$$
$$(11,6) = \{11, 6, 15, 9, 14, 5\}$$
$$(0,0) = \{0\}.$$

Therefore we have

$$XIB(1,4) = \{(1,14), (3,9), (16,7)\}$$
$$XIB(2,8) = \{(2,18), (6,8), (12,4)\}$$
$$XIB(4,11) = \{(4,1), (7,6), (9,3)\}$$
$$XIB(5,2) = \{(5,12), (17,5), (18,13)\}$$
$$XIB(8,1) = \{(8,11), (13,2), (19,17)\}$$
$$XIB(11,6) = \{(11,16), (15,19), (14,15)\}$$
$$XIB(0,0) = \{(0,10)\}.$$

Hence the presentation function of this N_{21} of type b) is:

x	0	1	a	a^2	a^3	a^4	a^5	a^6	a^7	a^8	a^9	a^{10}	a^{11}
T(x)	1	a^{10}	a^{14}	a^{18}	a^9	a	a^{12}	a^8	a^6	a^{11}	a^3	0	a^{16}

x	a^{12}	a^{13}	a^{14}	a^{15}	a^{16}	a^{17}	a^{18}	a^{19}
T(x)	a^4	a^2	a^{15}	a^{19}	a^7	a^5	a^{13}	a^{17} .

We have constructed two proper XIP-neofields, one with order 15 and one with order 21. These two examples are interesting because CIP-neofield does not exist for the order v = 15 or v = 21. In fact, Doner[4] has shown that CIP-neofield does not exist for any order v, $v \equiv 15, 21 \pmod{24}$.

Example I.62: Let v = 17, then $\ell = 8$. We have two sextuples and one triple in the XIP-admissible partition of type a) of Z_{16}^* .

$$(2, 7) = \{2, 7, 5, 14, 9, 11\}$$
$$(3, 4) = \{3, 4, 1, 13, 12, 15\}$$
$$(0, 6) = \{0, 6, 6, 0, 10, 10\}.$$

First of all, we have

$$SIB(2,7) = \{(2,15),(5,6),(9,3),(7,13),(14,1),(11,10)\}$$
$$SIB(3,4) = \{(3,12),(1,5),(12,7),(4,9),(13,4),(15,11)\}$$
$$SIB(0,6) = \{(0,14),(6,8),(10,2)\}$$

and the presenation function of a proper XIP-neofield:

x	0	1	a	a^2	a^3	a^4	a^5	a^6	a^7	a^8	a^9	a^{10}
T(x)	1	a^{14}	a^5	a^{15}	a^{12}	a^9	a^6	a^8	a^{13}	0	a^3	a^2

x	a^{11}	a^{12}	a^{13}	a^{14}	a^{15}
T(x)	a^{10}	a^7	a^4	a	a^{11}

This proper XIP-neofield is shown to be XIP-neofield of type a)ii). Secondly, we have

$$XIB(2,7) = SIB(2,7) = \{(2,15),(5,6),(9,3),(7,13),(14,1)\\ (11,10)\}$$

$$XIB(3,4) = CIB(3,4) = \{(3,12),(1,5),(12,7),(4,11),(13,9)\\ (15,4)\}$$

$$XIB(0,6) = \{(0,14),(6,8),(10,2)\},$$

and the presentation function:

x	0	1	a	a^2	a^3	a^4	a^5	a^6	a^7	a^8	a^9	a^{10}
T(x)	1	a^{14}	a^5	a^{15}	a^{12}	a^{11}	a^6	a^8	a^{13}	0	a^3	a^2

x	a^{11}	a^{12}	a^{13}	a^{14}	a^{15}
T(x)	a^{10}	a^7	a^9	a	a^4

This proper XIP-neofield is shown to be XIP-neofield of type a)iii). We will give the general account of this construction in Chapter II.

CHAPTER II

TYPE a) XIP-NEOFIELDS

Definition I.45 and Definition I.58 in Chapter I indicate that there are two types of XIP-admissible partitions which give rise to XIP-neofields. We will devote this chapter to the study of those XIP-neofields derived from type a) XIP-admissible partitions. Hence it is understood that those XIP-neofields are called type a) XIP-neofields. (see Appendix I)

In [4], Doner has shown that XIP-neofields of type a) i) (which have CIP property) cannot exist for certain finite orders (*viz*, $v \equiv 0, 6, 12, 15, 18, 21 \pmod{24}$ and 10), and has provided the constructions of CIP-neofields for all remaining finite orders.

Sec. 1. Additive Structure in SIP-Neofields

Definition II.1: The mapping *T** is defined as: for any cyclic neofield N_v and $x+y = z$ in N_v, then $T^*(N_v)$ has the property $z(y+x) = xy$, where $z \neq 0$, $x \neq 0$, $y \neq 0$.

Definition II.2: A cyclic neofield N_v is said to have *SP* if $T^*(N_v) = N_v$. That is:

$$N_v \text{ has SP} \Leftrightarrow (x+y)(y+x) = xy, \forall x,y \in N_v.$$

Lemma II.3: $(T^*)^2 = I$, the identity mapping.

Proof: Let N_v be a cyclic neofield with $x+y = z$. Hence in $T^*(N_v)$, we have $z(y+x) = xy$. Then in $T^{*2}(N_v)$, we have

$$xy \cdot (zx+zy) = zy \cdot zx.$$

Since multiplication is commutative and we have the distributive law, it follows that

$$xy \cdot z \cdot (x+y) = z^2 \cdot yx$$

i.e., $x+y = z$. Hence $(T^*)^2 = I$. q.e.d.

Definition II.4: A cyclic neofield N_v is said to be an SIP-neofield if N_v has XIP and SP.

Lemma II.5: If N_v is an SIP-neofield of order v with generator a, then

a) If v is even, any of the statements

i) $1+a^k = a^n$,

ii) $1+a^{n-k} = a^{-k}$,

iii) $1+a^{-k} = a^{-n}$,

iv) $1+a^{k-n} = a^k$,

v) $1+a^n = a^{n-k}$,

vi) $1+a^{-n} = a^{k-n}$,

implies the remaining five.

b) If v is odd, let $\ell = \frac{v-1}{2}$, then any of the following:

i) $1+a^k = a^{n+\ell}$

ii) $1+a^{n-k} = a^{-k+\ell}$

iii) $1+a^{-k} = a^{-n+\ell}$

iv) $1+a^{k-n} = a^{k+\ell}$,

v) $1+a^{n} = a^{n-k+\ell}$,

vi) $1+a^{-n} = a^{k-n+\ell}$,

implies the remaining five.

Proof: In both a) and b), each of i), ii), and vi) implies the remining two, this is proved in Lemma I.15. Analogous to this, each of iii), iv), and v) implies the remaining two. Hence it is sufficient to show that i) $\Rightarrow$ iii).

For v even, assuming i), we have, by SP,

$$\begin{aligned} 1+a^{k} = a^{n} &\Rightarrow a^{n}(a^{k}+1) = a^{k} \\ &\Rightarrow a^{n+k}(1+a^{-k}) = a^{k} \\ &\Rightarrow 1+a^{-k} = a^{-n}, \text{ this is iii).} \end{aligned}$$

For v odd, assuming i), we have, by SP,

$$\begin{aligned} 1+a^{k} = a^{n+\ell} &\Rightarrow a^{n+\ell}(a^{k}+1) = a^{k} \\ &\Rightarrow a^{n+k+\ell}(1+a^{-k}) = a^{k} \\ &\Rightarrow 1+a^{-k} = a^{-n-\ell} = a^{-n+\ell}, \text{ this is iii).} \end{aligned}$$

Since an SIP-neofield is also an XIP-neofield, Lemma I.35, Definition I.36, Definition I.37, and Lemma I.38 also hold for SIP-neofields.

Lemma II.6: Let N_v be any SIP-neofield of even order v. Let $1+a^{k} = a^{n}$ and $1+a^{s} = a^{t}$ in N_v. Let $S(k,n)=\{k, n, n-k, -k, -n, k-n\}$ for an admissible pair (k,n) as defined in Definition I.36. If $S(k,n) \cap S(s,t) \neq \emptyset$, then S(k,n) =

S(s,t). (We assume $|S(k,n)| = |S(s,t)| = 6$.)

Proof: Let $i \in S(k,n) \cap S(s,t)$ and assume $1+a^i = a^j$ in N_v. By a) of Lemma II.5, $1+a^i = a^j$ is one of the six additions equivalent to $1+a^k = a^n$, and the residues i,j occur in S(k,n) consecutively. Since any two consecutive elements of S(k,n) determine all of S(k,n), then

$$S(k,n) = S(i,j) = S(s,t).$$ q.e.d.

Corollary II.7: If there exists an SIP-neofield of even order v, then $v \equiv 2$ or 4 (mod 6).

Proof: By the above lemma, the set of all possible S(k,n)'s corresponding to the addition $1+a^k = a^n$ in N_v consists of disjoint sets all of which have cardinality 6, or all but one of which has cardinality 6, that exception having cardinality 2. These sets partition $\mathbb{Z}_{v-1}-\{0\}$, so $v \equiv 2, 4$ (mod 6). q.e.d.

It is clear from the above argument that for v even and N_v an SIP-neofield of order v, then $\{S(k,n) \mid 1+a^k = a^n\}$ is an XIP-admissible partition of type a). Moreover, Lemma II.5 a) shows that each SIP-neofield gives rise to defining relations with type a) ii) in Definition I.45 and Theorem I.46. Hence SIP-neofields are those XIP-neofields arising from type a) ii) XIP-admissible partition in the sense of Definition I.45.

Since a type a) XIP-admissible partition of

$\mathbb{Z}_{v-1}-\{0\}$ with $v \equiv 2$ or $4 \pmod 6$ is given in Appendix I at the end of this monograph, we only give two examples here to show how to construct SIP-neofields from a given XIP-admissible partition of type a).

<u>Example II.8</u>: For $v = 14$, g.c.d.$(3,v-1) = 1$ and $\mathbb{Z}_{13}-\{0\} = \{1, 2, 3, \ldots, 12\}$, an XIP-admissible partition of $\mathbb{Z}_{13}-\{0\}$ of type a) is given by

$$S(1,4) = \{1, 4, 3, 12, 9, 10\}$$
$$S(2,7) = \{2, 7, 5, 11, 6, 8\}.$$

We then have:

$$SIA(1,4) = \{(1,4), (3,12), (9,10), (4,3), (12,9), (10,1)\}$$
$$SIA(2,7) = \{(2,7), (5,11), (6,8), (7,5), (11,6), (8,2)\}$$

Hence the presentation function is:

x	0	1	a	a^2	a^3	a^4	a^5	a^6	a^7	a^8	a^9	a^{10}	a^{11}	a^{12}
$T(x)$	1	0	a^4	a^7	a^{12}	a^3	a^{11}	a^8	a^5	a^2	a^{10}	a	a^6	a^9

<u>Example II.9</u>: For $v = 16$, $v \equiv 4 \pmod 6$ and $3 \mid (v-1)$. An XIP-admissible partition of $\mathbb{Z}_{15}-\{0\}$ is given by:

$$S(1,4) = \{1, 4, 3, 14, 11, 12\}$$
$$S(2,8) = \{2, 8, 6, 13, 7, 9\}$$
$$S(5,10) = \{5, 10\}.$$

Then

$$SIA(1,4) = \{(1,4),\ (3,14),\ (11,12),\ (4,3),\ (14,11),\ (12,1)\}$$

$$SIA(2,8) = \{(2,8),\ (6,13),\ (7,9),\ (8,6),\ (13,7),\ (9,2)\}$$

$$SIA(5,10) = \{(5,10),\ (10,5)\}.$$

Hence we have the presentation function:

x	0	1	a	a^2	a^3	a^4	a^5	a^6	a^7	a^8	a^9	a^{10}	a^{11}	a^{12}	a^{13}	a^{14}
$T(x)$	1	0	a^4	a^8	a^{14}	a^3	a^{10}	a^{13}	a^9	a^6	a^2	a^5	a^{12}	a	a^7	a^{11}

We now proceed to the case of an SIP-neofield of odd order v, developing from b) of Lemma II.5 an analysis of the additive structure for these SIP-neofields analogous to the development presented before for even order SIP-neofields. Again certain conditions (as in Definition I.53) on n and k, for the relation in N_v, $1+a^k = a^{n+\ell}$, $\ell = \frac{v-1}{2}$, guarantee that the six additions described in Lemma II.5,,b) are distinct. Examination of the cases arising when these conditions are relaxed leads to a complete characterization of the additive structure of odd-order SIP-neofields. Now we selectively relax any one of the conditions described in Definition I.53:

$$n \not\equiv 0 \pmod{v-1},\ k \not\equiv 0 \pmod{v-1},\ \text{and } k \not\equiv \pm n \pmod{v-1}.$$

<u>Case 1</u>: $n \equiv 0 \pmod{v-1}$.

From additions (v) and (vi) of Lemma II.5, we have

$$a^{n-k+\ell} = 1+a^n = 1+a^{-n} = a^{k-n+\ell}.$$

Hence $2k \equiv 0 \pmod{v-1}$. Then $k \equiv 0 \pmod{v-1}$ or $k \equiv \ell$

(mod v-1). But the latter case is impossible by Lemma I.14. Hence $k \equiv n \equiv 0 \pmod{v-1}$. We then have the singleton $SIB(0,0) = \{(0,\ell)\}$. Note that this happens only when $1+1 = a^{\ell} = -1 \pmod{v-1}$, hence $v \equiv 0 \pmod 3$ and then $v \equiv 3 \pmod 6$.

Case 2: $k \equiv 0 \pmod{v-1}$.

In this case, we have $S(0,n) = \{0, n, n, 0, -n, -n\}$, where $n \notin \{0,\ell\}$. It follows that

$$SIB(0,n) = \{(0,n+\ell), (n,\ell), (-n,-n+\ell), (n,n+\ell), (0,-n+\ell), (-n,\ell)\}.$$

By i) and iii) of Lemma II.5, b), we then have

$$a^{n+\ell} = 1+a^{k} = 1+a^{-k} = a^{-n+\ell},$$

which implies $n \equiv 0 \pmod{v-1}$; this reduces to Case 1.

Case 3: $k \equiv -n \pmod{v-1}$.

By Lemma II.5, b) i), ii), we get $n \equiv k-n \pmod{v-1}$. Then $2n \equiv k \equiv -n \pmod{v-1}$, which is: $3n \equiv 0 \pmod{v-1}$. If g.c.d.$(3,v-1) = 1$, then $n \equiv 0 \equiv k \pmod{v-1}$, and we get the singleton $SIB(0,0) = \{(0,\ell)\}$. If $3\,|\,(v-1)$, then $n = \frac{v-1}{3}$ and hence $k = \frac{2(v-1)}{3}$; we then have the pair

$$SIB\left(\frac{v-1}{3}, \frac{2(v-1)}{3}\right) = \left\{\left(\frac{v-1}{3}, \frac{v-1}{6}\right), \left(\frac{2(v-1)}{3}, \frac{5(v-1)}{6}\right)\right\}.$$

Hence the singleton and the pair are mutually exclusive. The latter case occurs when $3\,|\,(v-1)$, i.e., $v \equiv 1 \pmod 6$.

Case 4: $k \equiv n \pmod{v-1}$.

Using i) and v) of Lemma II.5, b), we get $n \equiv n-k \pmod{v-1}$. Hence $k \equiv 0 \pmod{v-1}$. This would reduce to Case 2, hence we get a singleton $SIB(0,0) = \{(0,\ell)\}$ when $v \equiv 3 \pmod 6$.

Hence we have the following lemma:

<u>Lemma II.10</u>: If N_v is an SIP-neofield of odd order v, and $1+a^k = a^{n+\ell}$, $k,n \in \mathbb{Z}_{v-1}$, then $|SIB(k,n)| = 6, 2$ or 1.

<u>Theorem II.11</u>: If an SIP-neofield N_v of odd order v exists then $v \equiv 3 \pmod 6$, and the sets SIB(k,n), determined by additions $1+a^k = a^{n+\ell}$ in N_v, are all of cardinality 6, except the singleton $SIB(0,0) = \{(0,\ell)\}$ where $\ell = \frac{v-1}{2}$.

<u>Proof</u>: Similar to the proof of Lemma II.7 in the even case, we have: if $1+a^k = a^{n+\ell}$ and $1+a^r = a^{s+\ell}$ in the SIP-neofield N_v, then SIB(k,n) and SIB(r,s) are either disjoint or equal (mod v-1), where

$$SIB(k,n) = \{(k,n+\ell),\ (n,n-k+\ell),\ (n-k,-k+\ell),\ (-k,-n+\ell),\ (-n,k-n+\ell),\ (k-n,k+\ell)\}.$$

The set of all possible SIB(k,n)'s corresponding to the addition $1+a^k = a^{n+\ell}$ in N_v consists of sets: (1) all of which have cardinality 6, (2) all of which have cardinality 6, except one which has cardinality 2, (3) all of which have cardinality 6, except one which has cardinality 1, or (4) all of which have cardinality 6 except one which has cardinality 2 and one which has cardinality 1.

We will count the exponents $n+\ell$ of elements $a^{n+\ell}$

in N_v which occur in sums $1+a^k = a^{n+\ell}$ in N_v; those exponents occur as right hand entries of the ordered pairs in the disjoint sets in $C = \{SIB(k,n) \mid 1+a^k = a^{n+\ell} \text{ in } N_v\}$. Two elements never occur in the role of $a^{n+\ell}$, i.e., 1 and 0.

If (1) holds, then $v-2 \equiv 0 \pmod 6$, a contradiction since v is odd. If (2) holds, then $v-2-2 \equiv 0 \pmod 6$, therefore $v \equiv 4 \pmod 6$, again a contradiction. If (4) holds, then $v-2-2-1 \equiv 0 \pmod 6$, i.e., $v \equiv 5 \pmod 6$. But since we have one SIB(k,n) with cardinality 2 by the case3 discussion preceding Lemma II.10, we have $v-1 \equiv 0 \pmod 6$, i.e., $v \equiv 1 \pmod 6$, which contradicts $v \equiv 5 \pmod 6$. Then we have the only possible case (3), which shows that $v-2-1 \equiv 0 \pmod 6$. Hence $v \equiv 3 \pmod 6$. And the sets SIB(k,n), determined by additions $1+a^k = a^{n+\ell}$ in N_v, are all of cardinality 6, except for the singleton $SIB(0,0) = \{(0,\ell)\}$. q.e.d.

Corollary II.12: There doesn't exist an SIP-neofield of order $v \equiv 1$ or $5 \pmod 6$.

It is clear that for $v \equiv 3 \pmod 6$ and N_v an SIP-neofield of order v, then $\{S(k,n) \mid 1+a^k = a^{n+\ell}\}$ is an XIP-admissible partition of $\mathbb{Z}_{v-1} - \{\ell\}$ of type a). Moreover, Lemma II.5 b) shows that SIP-neofield gives rise to defining relations with type a) ii) in Definition I.58 and Theorem I.59. Hence the SIP-neofields of order v (odd) are those XIP-neofields arising from type a) ii) XIP-admissible partitions, in the sense of Definition I.58.

Since a type a) XIP-admissible partition of

$\mathbb{Z}_{v-1}-\{\ell\}$ with $v \equiv 3 \pmod 6$ is listed in Appendix I at the end of this monograph, we only present two examples here to show how to construct SIP-neofields from a given XIP-admissible partition of type a).

Example II.13: For $v = 9$, $\ell = 4$, an XIP-admissible partition of type a) of $\mathbb{Z}_8-\{4\}$ is given by

$$S(1,6) = \{1, 6, 5, 7, 2, 3\}$$
$$S(0,0) = \{0\}.$$

Hence

$$SIB(1,6) = \{(1,2), (6,1), (5,3), (7,6), (2,7), (3,5)\}$$
$$SIB(0,0) = \{(0,4)\}.$$

We have the presentation function of an SIP-neofield:

x	0	1	a	a^2	a^3	a^4	a^5	a^6	a^7
$T(x)$	1	a^4	a^2	a^7	a^5	0	a^3	a	a^6.

Example II.14: For $v = 27$, $\ell = 13$, an XIP-admissible partition of type a) of $\mathbb{Z}_{26}-\{13\}$ is given by

$$S(1,4) = \{1, 4, 3, 25, 22, 23\}$$
$$S(2,9) = \{2, 9, 7, 24, 17, 19\}$$
$$S(5,15) = \{5, 15, 10, 21, 11, 16\}$$
$$S(6,14) = \{6, 14, 8, 20, 12, 18\}$$
$$S(0,0) = \{0\}.$$

Hence

$SIB(1,4) = \{(1,17), (4,16), (3,12), (25,9), (22,10), (23,14)\}$

$SIB(2,9) = \{(2,22), (9,15), (7,11), (24,4), (17,6), (19,15)\}$

$SIB(5,15) = \{(5,2), (15,23), (10,8), (21,24), (11,3), (16,18)\}$

$SIB(6,14) = \{(6,1), (14,21), (8,7), (20,25), (12,5), (18,19)\}$

$SIB(0,0) = \{(0,13)\}$.

Therefore we have the presentation function of an SIP, N_{27}:

x	0	1	a	a^2	a^3	a^4	a^5	a^6	a^7	a^8	a^9	a^{10}	a^{11}	a^{12}
$T(x)$	1	a^{13}	a^{17}	a^{22}	a^{12}	a^{16}	a^2	a	a^{11}	a^7	a^{20}	a^8	a^3	a^5

x	a^{13}	a^{14}	a^{15}	a^{16}	a^{17}	a^{18}	a^{19}	a^{20}	a^{21}	a^{22}	a^{23}	a^{24}	a^{25}
$T(x)$	0	a^{21}	a^{23}	a^{18}	a^6	a^9	a^{15}	a^{25}	a^{24}	a^{10}	a^{14}	a^4	a^9 .

Sec. 2. Non-Existence Theorems and the Constructions.

Theorem II.15: There does not exist an SIP-neofield of order $v \equiv 15$ or $21 \pmod{24}$.

Proof: We will count even residues of $\mathbb{Z}_{v-1}-\{0,\ell\}$ in two different ways and derive a contradition.

Let $v = 24m+6i+3$, $i = 2$, or 3. If there exists SIP-neofield of order $v \equiv 15$ or $21 \pmod{24}$, then $v \equiv 3 \pmod 6$. By Theorem II.11, the sets $SIB(k,n)$ determined by additions $1+a^k = a^{n+\ell}$ in N_v are all of cardinality 6, except for the singleton $SIB(0,0) = \{(0,\ell)\}$. Hence, there is a partition Π of $\mathbb{Z}_{v-1}-\{0,\ell\}$ into $\frac{v-3}{6} = 4m+i$ sextuples of the form

$$S(k,n) = \{k, n, n-k, -k, -n, k-n\}$$

with all $|S(k,n)| = 6$. Since v is odd, v-1 is even. It follows that $\mathbb{Z}_{v-1}-\{0,\ell\} = \{1, 2, 3, \ldots, 24m+6i+1\} - \{12m+3i+1\}$ contains

$$\begin{cases} \frac{v-1}{2}+1-2 = \frac{v-3}{2}, \text{ when } i = 2 \\ \frac{v-1}{2}-2 = \frac{v-5}{2}, \text{ when } i = 3 \end{cases}$$

even residues, i.e., contains $\begin{cases} 12m+3i \\ 12m+3i-1 \end{cases}$ even residues for $\begin{cases} i=2 \\ i=3 \end{cases}$. If k,n are all even, then every element in S(k,n) is even. Otherwise the number of even residues in S(k,n) is 2. Suppose π is a partition of $\mathbb{Z}_{v-1}-\{0,\ell\}$ into t sets of $S(k_i,n_i)$ with k_i,n_i even, i = 1, 2, ..., t, then the number of even residues is $6t+2\cdot(4m+i-t) = 8m+4t+2i = \begin{cases} 12m+3i, \text{ when} \\ 12m+3i, \text{ when} \end{cases}$ $\begin{matrix} i=2 \\ i=3 \end{matrix}$. Hence $4t-4m = \begin{cases} i, \text{ if } i = 2 \\ i-1, \text{ if } i = 3 \end{cases} = 2$ and then we have $4|2$, which is absurd. Therefore there is no SIP-neofield of order $v \equiv 15$ or 21 (mod 24). q.e.d.

Lemma II.16: There does not exist an SIP-neofield of order 10.

Proof: When v = 10, $\mathbb{Z}_{v-1}-\{0\} = \{1, 2, 3, 4, 5, 6, 7, 8\}$. By Lemma I.35, we should have an XIP-admissible partition of type a), i.e., a sextuple S(k,n) for $k,n \in \{1, 2, 4, 5, 7, 8\}$ and a pair $\{3,6\}$. Check all the possibilities exhaustively to find k,n in $\{1, 2, 4, 5, 7, 8\}$ such that (k,n) generate S(k,n), we claim that this can never be done. Hence there is no SIP-neofield of order 10. q.e.d.

<u>Theorem II.17</u>: For $v \equiv 0, 1, 5, 6, 7, 11, 12, 13, 15, 17, 18, 19, 21, 23 \pmod{24}$ and $v = 10$, no SIP-neofield of order v exists. SIP-neofields exist for all other orders.

<u>Proof</u>: By Corollary II.7 and II.12, $v \equiv 0, 6, 12, 18, 1, 5, 7, 11, 13, 17, 19, 23 \pmod{24}$ are forbidden for the existence of an SIP-neofield. These arguments, together with Theorem II.15 and Lemma II.16, yield the first part of the theorem. By Theorem I.46, a SIP-neofield of order $v \equiv 2, 4 \pmod 6$, $v \neq 10$ exists if there is an XIP-admissible partition of $\mathbf{Z}_{v-1}-\{0\}$ of type a). By Theorem I.59, SIP-neofield of order $v \equiv 3 \pmod 6$ exists if there is an XIP-admissible partition of $\mathbf{Z}_{v-1}-\{\ell\}$ of type a). Since the construction of XIP-admissible partitions for $v \equiv 2, 3, 4 \pmod 6$ of type a) except $v \equiv 15, 21 \pmod{24}$ is provided in Appendix I, the theorem follows from the above discussion. <u>q.e.d.</u>

We end this section with a theorem dealing with the intersection between the families of SIP- and CIP-neofields. Note that we don't assume the associativity of the addition.

<u>Theorem II.18</u>: There doesn't exist any cyclic neofield of order $v > 4$ which has both CIP and SIP.

<u>Proof</u>: Suppose the cyclic neofield N_v of order v has both SIP and CIP. For any $k \in \begin{cases} \mathbf{Z}_{v-1}-\{0\}, & v \text{ even} \\ \mathbf{Z}_{v-1}-\{\ell\}, & v \text{ odd} \end{cases}$, let

$$1+a^k = \begin{cases} a^n, & v \text{ even} \\ a^{n+\ell}, & v \text{ odd} \end{cases}.$$

If v is even, then by Lemma II.5, a), we have $1+a^n = a^{n-k}$ and by Lemma I.3 in [4], we have $1+a^n = a^k$. Hence $n-k \equiv k \pmod{v-1}$, i.e., $n \equiv 2k \pmod{v-1}$. Therefore $S(k,n) = \{k, 2k, k, -k, -2k, -k\}$. Apply Lemma II.5, a) and Lemma I.3 again, we have

$$a^{-2k} = 1+a^{-k} = a^k, \text{ i.e., } 3k \equiv 0 \pmod{v-1}.$$

If g.c.d.(3,v-1) = 1, then $k \equiv 0 \pmod{v-1}$, a contradiction to Lemma I.17. Hence $3 \mid (v-1)$ and $k = \frac{v-1}{3}$, $n = \frac{2(v-1)}{3}$. Since k is arbitrary, we have

$$N_v = \{0, 1, a^{(v-1)/3}, a^{2(v-1)/3}\}.$$

Hence $N_v = \{0, 1, b, b^2\}$ with $1+b = b^2$, $1+b^2 = b$, which is the field $\mathbb{F}_4$.

If v is odd, the preceding paragraph goes through, (except that $1+x = y$ is replaced by $1+x = ya^{\ell}$) until we get $3k \equiv 0 \pmod{v-1}$. Since v-1 is even, we have either $k \equiv 0 \pmod{v-1}$ or $k = \frac{v-1}{3}$. In the first case, it follows from Case 2 in the discussion preceding Lemma II.10 that $n \equiv 0 \pmod{v-1}$, hence SIB(0,n) is reduced to singleton, contradicts to $v > 4$. In the second case, we have by Case 3 in the discussion preceding Lemma II.10 that

$$N_v = \{0, 1, a^{(v-1)/6}, a^{(v-1)/3}, a^{2(v-1)/3}, a^{5(v-1)/6}\},$$

this is absurd since v is odd.

Hence there is no cyclic neofield of order >4 which has both SIP and CIP. q.e.d.

Sec. 3. Construction of Type a) iii) XIP-Neofield.

In Chapter I, an XIP-neofield is called proper if it is not a CIP-neofield. Now, an XIP-neofield will be called pure proper if it is neither a CIP- nor an SIP-neofield. It was shown in Theorem I.41 that for $v \equiv 0 \pmod 6$, no pure proper XIP-neofield of order v exists. For any other finite v, except for $v \equiv 15, 21 \pmod{24}$, the existence of a pure proper XIP-neofield of type a) is provided by Appendix I. Since this type of pure proper XIP neofields arise from type a) XIP-admissible partition, we call them type a) iii) XIP-neofields.

Since by [4], no XIP-admissible partition of type a) exists for $v \equiv 15$ or $21 \pmod{24}$, a pure proper XIP-neofield of type a) iii) doesn't exist.

Definition II.19: An XIP-neofield is called pure proper if it has neither CIP nor SIP.

For v even, by Definition I.45, Theorem I.46, an XIP-admissible partition of $\mathbb{Z}_{v-1} - \{0\}$ of type a) iii) would give rise to an XIP-neofield. We will show that those XIP-neofields arising from XIP-admissible partiton of type a) iii) are pure proper and give some examples.

Lemma II.20: XIP-neofields derived from type a) iii) XIP-admissible partitions of $\mathbf{Z}_{v-1}-\{0\}$ have neither CIP nor SIP.

Proof: In order to have a CIP-neofield, we should have $XIA(k,n) = CIA(k,n)$ for all (k,n) of the sextuple $S(k,n)$ in a) i) of Definition I.45. Hence the XIP-neofield N_v arising from a) iii) of Definition I.45 is not CIP. Similarly, N_v doesn't have SIP. Hence N_v has neither SIP nor CIP. q.e.d.

Example II.21: For $v = 14$, $v \equiv 2 \pmod 6$. An XIP-admissible partition of type a) of $\mathbf{Z}_{13}-\{0\}$ is:

$$(3,4) = \{3, 4, 1, 10, 9, 12\}$$
$$(5,7) = \{5, 7, 2, 8, 6, 11\}.$$

Then

$$XIA(3,4) = SIA(3,4) = \{(3,4), (1,10), (9,12), (4,1), (10,9), (12,3)\}$$
$$XIA(5,7) = CIA(5,7) = \{(5,7), (2,8), (6,11), (7,5), (8,2), (11,6)\}.$$

Hence we get a pure proper XIP, N_{14} with the presentation function:

x	0	1	a	a^2	a^3	a^4	a^5	a^6	a^7	a^8	a^9	a^{10}	a^{11}	a^{12}
$T(x)$	1	0	a^{10}	a^8	a^4	a	a^7	a^{11}	a^5	a^2	a^{12}	a^9	a^6	a^3

Example II.22: For $v = 22$, $v \equiv 4 \pmod 6$. An XIP-admissible partition of $\mathbf{Z}_{21}-\{0\}$ of type a) is given by

$$(4,5) = \{4,\ 5,\ 1,\ 17,\ 16,\ 20\}$$

$$(8,10) = \{8,\ 10,\ 2,\ 13,\ 11,\ 19\}$$

$$(6,9) = \{6,\ 9,\ 3,\ 15,\ 12,\ 18\}$$

$$(7,14) = \{7,\ 14\}.$$

Then we take

$$XIA(4,5) = CIA(4,5) = \{(4,5),\ (1,17),\ (16,20),\ (5,4),\ (17,1),\ (20,16)\}$$

$$XIA(8,10) = SIA(8,10) = \{(8,10),\ (2,13),\ (11,19),\ (10,2),\ (13,11),\ (19,8)\}$$

$$XIA(6,9) = SIA(6,9) = \{(6,9),\ (3,15),\ (12,18),\ (9,3),\ (15,12),\ (18,6)\}$$

$$XIA(7,14) = \{(7,14),\ (14,7)\}$$

Hence we have the following pure proper XIP-neofield of type a) iii):

x	0	1	a	a^2	a^3	a^4	a^5	a^6	a^7	a^8	a^9	a^{10}	a^{11}	a^{12}	a^{13}
$T(x)$	1	0	a^{17}	a^{13}	a^{15}	a^5	a^4	a^9	a^{14}	a^{10}	a^3	a^2	a^{19}	a^{18}	a^{11}

x	a^{14}	a^{15}	a^{16}	a^{17}	a^{18}	a^{19}	a^{20}
$T(x)$	a^7	a^{12}	a^{20}	a	a^6	a^8	a^{16}.

<u>Remark II.23</u>: If N_v is a pure proper XIP-neofield of even order v, associated with which is an XIP-admissible partition of $\mathbb{Z}_{v-1}-\{0\}$ which contains $S(k,n) = \{k,\ n,\ n-k,\ -k,\ -n,\ k-n\}$, then either $1+a^k = a^n$ or $1+a^n = a^k$. Thus we have $2^{[v/6]}$* choices. Moreover, XIA(k,n) is defined as either CIA(k,n) or SIA(k,n). Since $XIA(k,n) \not\equiv CIA(k,n), \not\equiv SIA(k,n)$, for an XIP-admissible partition, we then have

*[z] is the greatest integer less than or equal to z.

$2^{[v/6]}-2$ possibilities to define XIA(k,n). Hence an XIP-admissible partition of $\mathbf{Z}_{v-1}-\{0\}$ of type a) doesn't uniquely determine a pure proper XIP-neofield of order v. In fact, there are $2^{[v/6]}(2^{[v/6]}-2)$ proper XIP-neofields of type a) iii) arising from an XIP-admissible partition of $\mathbf{Z}_{v-1}-\{0\}$, for $v > 14$.

We now turn our attention to the case when v is odd. As before, we let $\ell = \frac{v-1}{2}$.

It is proved in Lemma I.57 that for odd v and N_v an XIP-neofield of order v, $\{S(k,n) | 1+a^k = a^{n+\ell}\}$ is an XIP-admissible partition of $\mathbf{Z}_{v-1}-\{0\}$ of either type a) or type b) defined in Lemma I.55. The converse part of this is proved in Theorem I.59. Here we are only interested in the problem of deriving a pure proper XIP-neofield of type a) iii) from a given XIP-admissible partition of type a).

For $v \equiv 3 \pmod 6$, except $v \equiv 15$ or $21 \pmod{24}$, we have an SIP-neofield of order v (by Theorem II.17) and a CIP-neofield of order v (by [4]). Hence we are able to derive pure proper XIP-neofields by using a) iii) of Definition I.58.

For $v \equiv 1$ or $5 \pmod 6$, CIP-neofields do exist (by Doner [4]). But SIP-neofield doesn't. By Theorem II.11 and the discussion before Lemma II.10, when $k \equiv 0 \pmod{v-1}$, the sextuple S(0,n) becomes

$$S(0,n) = \{0,\ n,\ n,\ 0,\ -n,\ -n\},\ n \notin \{0, \ell\}.$$

It follows that

$$SIB(0,n) = \{(0,n+\ell), (n,\ell), (-n,-n+\ell), (n,n+\ell), (0,-n+\ell), (-n,\ell)\}.$$

But $(0,n+\ell)$ and $(0,-n+\ell)$ are mutually exclusive in SIB(0,n) unless $n \equiv 0 \pmod 6$; then SIB(0,n) reduces to the singleton $SIB(0,0) = \{(0,\ell)\}$. The crucial point is that we don't really need SIB(0,n), we only need XIB(0,n) as follows:

$$SIB(0,n) = \{(0,n+\ell), (n,\ell), (-n,-n+\ell)\}.$$

Hence for any other (k,n) such that S(k,n) is a sextuple, we define XIB(k,n) as in a) iii) of Definition I.58. We still get pure proper XIP-neofields. We give three examples to illustrate each of the three cases $v \equiv 1$, 3 and 5 (mod 6).

<u>Example II.24</u>: $v = 13, \mathbf{Z}_{12}-\{\ell\} = \{0, 1, 2, \ldots, 5, 7, 8, \ldots, 11\}$. An XIP-admissible partition of $\mathbf{Z}_{12}-\{\ell\}$ of type a) is given as:

$$(1,3) = \{1, 3, 2, 11, 9, 10\}$$
$$(0,5) = \{0, 5, 5, 0, 7, 7\}$$
$$(4,8) = \{4,8\}.$$

Then we have:

$$XIB(1,3) = SIB(1,3) = \{(1,9), (2,5), (9,4), (3,8), (11,3), (10,7)\}$$
$$XIB(0,5) = \{(0,11), (5,6), (7,1)\}$$
$$XIB(4,8) = \{(4,2), (8,10)\}.$$

Hence we have a pure proper XIP-neofield N_{13} of type a) iii).

x	0	1	a	a^2	a^3	a^4	a^5	a^6	a^7	a^8	a^9	a^{10}	a^{11}
$T(x)$	1	a^{11}	a^9	a^5	a^8	a^2	a^6	0	a	a^{10}	a^4	a^7	a^3 .

<u>Example II.25</u>: For $v = 27$, $\ell = 13$, $v \equiv 3 \pmod 6$, an XIP-admissible partition of type a) is given by:

$$(1,4) = \{1, 4, 3, 25, 22, 23\}$$
$$(2,9) = \{2, 9, 7, 24, 17, 19\}$$
$$(5,15) = \{5, 15, 10, 21, 11, 16\}$$
$$(6,14) = \{6, 14, 8, 20, 12, 18\}$$
$$(0,0) = \{0\}.$$

Then take

$$XIB(1,4) = SIB(1,4) = \{(1,17), (4,16), (3,12), (25,9), (22,10), (23,14)\}$$
$$XIB(2,9) = CIB(2,9) = \{(2,22), (9,15), (7,11), (24,20), (17,6), (19,4)\}$$
$$XIB(5,15) = SIB(5,15) = \{(5,2), (15,23), (10,8), (21,24), (11,3), (16,18)\}$$
$$XIB(6,14) = CIB(6,14) = \{(6,1), (14,19), (8,7), (20,21), (12,5), (18,25)\}$$
$$XIB(0,0) = \{(0,\ell)\} = \{(0,13)\}.$$

Heence we have the presentation function of a pure proper XIP-neofield N_{27} of type a) iii).

x	0	1	a	a^2	a^3	a^4	a^5	a^6	a^7	a^8	a^9	a^{10}	a^{11}	a^{12}
$T(x)$	1	a^{13}	a^{17}	a^{22}	a^{12}	a^{16}	a^2	a	a^{11}	a^{17}	a^{15}	a^8	a^3	a^5

x	a^{13}	a^{14}	a^{15}	a^{16}	a^{17}	a^{18}	a^{19}	a^{20}	a^{21}	a^{22}	a^{23}	a^{24}	a^{25}
$T(x)$	0	a^{19}	a^{23}	a^{18}	a^6	a^{25}	a^4	a^{21}	a^{24}	a^{10}	a^{14}	a^{20}	a^9 .

Example II.26: For $v \equiv 5 \pmod 6$, take $v = 11$. We have one sextuple and one triple in the XIP-admissible partition of type a) of $\mathbf{Z}_{10} - \{5\}$.

$$(3,4) = \{3, 4, 1, 7, 6, 9\}$$
$$(0,2) = \{0, 2, 2, 0, 8, 8\}.$$

Hence in order to get a pure proper XIP-neofield of type a) iii), we are forced to take

$$XIB(3,4) = SIB(3,4) = \{(3,9), (4,6), (1,2), (7,1), (6,4), (9,8)\}$$
$$XIB(0,2) = \{(0,7), (2,5), (8,3)\},$$

then the presentation function is:

x	0	1	a	a^2	a^3	a^4	a^5	a^6	a^7	a^8	a^9
$T(x)$	1	a^7	a^2	a^5	a^9	a^6	0	a^4	a	a^3	a^8 .

Remark II.27: For v odd, $v \geq 11$, we have a pure proper XIP-neofield of type a) iii) except for the forbidden orders $v \equiv 15, 21 \pmod{24}$. For $v \equiv 15, 21 \pmod{24}$, there doesn't exist any XIP-admissible partition of type a) as shown in Theorem II.15. But pure proper XIP-neofields do exist for $v \equiv 15, 21 \pmod{24}$ as shown in Examples I.60, I,61. In fact, those pure proper XIP-neofields give rise to

XIP-admissible partitions of type b). We will give a general construction of XIP-neofields of type b) in Chapter III.

CHAPTER III

CONSTRUCTION OF TYPE b) XIP-NEOFIELDS

Sec. 1. Construction of Even Order Type b) XIP-Neofields

In this section, constructions and examples are provided verifying the existence of even order XIP-neofields of type b) (in the sence of Definition I.43) of orders $v \equiv 2 \pmod 6$ and $v \equiv 4 \pmod{18}$. Recalling Theorem I.41, we have that no XIP-neofield N_v of orders $v \equiv 0 \pmod 6$ exists; in addition, no XIP-neofield of order 10 exists, as can be verified by attempting to construct one using Theorem I.46.

Case 1: $v = 6m+2$, $m \geq 2$.

We partition Z^*_{6m+1} into 2m sextuples $\{k, n, n-k, -k, -n, k-n\}$ which give a type b) XIP-admissible partition of Z^*_{6m+1}.

First, pair k and n by the following scheme EA1

k	n	n-k	-k	-n	k-n
1	m+2	m+1	6m	5m-1	5m
2	m+4	m+2	6m-1	5m-3	5m-1
3	m+6	m+3	6m-2	5m-5	5m-2
.	.	.	.	.	.
.	.	.	.	.	.

k	n	n-k	-k	-n	k-n
.	.	.	.	.	.
m-1	3m-2	2m-1	5m+2	3m+3	4m+2
m	3m	2m	5m	31+1	4m+1

EA1

yielding by a "nesting" procedure of the odd-parity elements

$$\{k,\ n-k,\ -n\}$$

for each odd element from 3m+1 to 5m-1 when m is even (for each even element from 3m+1 to 5m-1 when m is odd) and all elements from 1 to 2m; and even-parity elements

$$\{n,\ -k,\ k-n\}$$

for each even element (odd element) from m+2 to 3m when m is even (when m is odd) and all elements from 4m+1 to 6m.

Then we pick k and n by the following scheme EB1.

k	n	n-k	-k	-n	k-n
3m	1	3m+2	3m+1	6m	3m-1
3m-1	2	3m+4	3m+2	6m-1	3m-3
3m-2	3	3m+6	3m+3	6m-2	3m-5
.	.	.	.	.	.
.	.	.	.	.	.
.	.	.	.	.	.
2m+2	m-1	5m-2	4m-1	5m+2	m+3

$\underline{k}$	$\underline{n}$	$\underline{n-k}$	$\underline{-k}$	$\underline{-n}$	$\underline{k-n}$
2m+1	m	5m	4m	5m	m+1

<u>EB1</u>

yielding by a "nesting" procedure odd-parity elements

$$\{k,\ n-k,\ -n\}$$

for each even element (odd element) from 3m+2 to 5m when m is even (whem m is odd) and all elements from 2m+1 to 3m and from 5m+1 to 6m; and even-parity elements

$$\{n,\ -k,\ k-n\}$$

for each odd element (even element) from m+1 to 3m-1 when m is even (when m is odd) and all elements from 1 to m and from 3m+1 to 4m.

One checks that all the odd-parity elements (all the even-parity elements) in schemes EA1 and EB1 range over all non-zero elements in $\mathbb{Z}_{6m+1}$. Moreover, one checks that $|S(1,m+2)\cap S(3m,1)| = 2$, so the partition is not of type a) in the sense of Definition I.43. Hence we have an XIP-admissible partition of $\mathbb{Z}^*_{6m+1}$ of type b), and by Definition I.45 b) and Theorem I.46, this construction determines an XIP-neofield of order v = 6m+2, m $\geq$ 2.

This construction requires that m $\geq$ 2: for m = 1, we have v = 8. There are only four XIP-neofields of order 8, two of them have CIP and the other two have SIP. Hence there doesn't exist a XIP-neofield with order 8 of type b).

Examples:

We provide 3 consecutive cases of this construction in the format used in the above construction.

For $v = 14$, $m = 2$, we have 4 sextuples:

$$S(1,4) = \{1, 4, 3, 12, 9, 10\} \quad S(6,1) = \{6, 1, 8, 7, 12, 5\}$$
$$S(2,6) = \{2, 6, 4, 11, 7, 9\} \quad S(5,2) = \{5, 2, 10, 8, 11, 3\}.$$

For $v = 20$, $m = 3$, 6 sextuples are given by

$$S(1,5) = \{1, 5, 4, 18, 14, 15\}$$
$$S(2,7) = \{2, 7, 5, 17, 12, 14\}$$
$$S(3,9) = \{3, 9, 6, 16, 10, 13\}$$
$$S(9,1) = \{9, 1, 11, 10, 18, 8\}$$
$$S(8,2) = \{8, 2, 13, 11, 17, 6\}$$
$$S(7,3) = \{7, 3, 15, 12, 16, 4\}.$$

For $v = 26$, $m = 4$, 8 sextuples are given by

$$S(1,6) = \{1, 6, 5, 24, 19, 20\}$$
$$S(2,8) = \{2, 8, 6, 23, 17, 19\}$$
$$S(3,10) = \{3, 10, 7, 22, 15, 18\}$$
$$S(4,12) = \{4, 12, 8, 21, 13, 17\}$$
$$S(12,1) = \{12, 1, 14, 13, 24, 11\}$$
$$S(11,2) = \{11, 2, 16, 14, 23, 9\}$$
$$S(10,3) = \{10, 3, 18, 15, 22, 7\}$$
$$S(9,4) = \{9, 4, 20, 16, 21, 5\}.$$

From the above example with $v = 26$, we have $S(3,10) = S(10,3)$, in the construction with the same sextuple appearing in both schemes EA1 and EB1. In general, by taking $s = \frac{v-8}{9}+1$ when $v = 6m+2$ and $m = 3f+1$, we have $v = 18f+8$ and $s = 2f+1$. Since all the sextuples $S(k,n)$ in scheme EA1 have the property that $2k+m = n$, by taking

$$t = 2s+m = 2(2f+1)+(3f+1) = 7f+3,$$

we have $S(s,t) = \{2f+1, 7f+3, 5f+2, 16f+7, 11f+5, 13f+6\}$ in scheme EA1. For sextuples in scheme EB1, we have $k+n = 3m+1$. By checking that

$$s+t = (2f+1)+(7f+3) = 3m+1,$$

we have $S(t,s)$ also in scheme EB1. To illustrate this, we provide one more example.

For $v = 44$, $m = 7$, $f = 2$. We have $S(5,17)$ in scheme EA1 and $S(17,5)$ in scheme EB1 for the following construction.

$$S(1,9) = \{1, 9, 8, 42, 34, 35\}$$
$$S(2,11) = \{2, 11, 9, 41, 32, 34\}$$
$$S(3,13) = \{3, 13, 10, 40, 30, 33\}$$
$$S(4,15) = \{4, 15, 11, 39, 28, 32\}$$
$$S(5,17) = \{5, 17, 12, 38, 26, 31\}$$
$$S(6,19) = \{6, 19, 13, 37, 24, 30\}$$
$$S(7,21) = \{7, 21, 14, 36, 22, 29\}$$
$$S(21,1) = \{21, 1, 23, 22, 42, 20\}$$
$$S(20,2) = \{20, 2, 25, 23, 41, 18\}$$
$$S(19,3) = \{19, 3, 27, 24, 40, 16\}$$

$$S(18,4) = \{18, 4, 29, 25, 39, 14\}$$

$$S(17,5) = \{17, 5, 31, 26, 38, 12\}$$

$$S(16,6) = \{16, 6, 33, 27, 37, 10\}$$

$$S(15,7) = \{15, 7, 35, 28, 36, 8\}.$$

By Definition I.45 b) and Theorem I.46, we have a type b) XIP-neofield N_{44} with $t = 1$.

Case 2: $v = 18m+4$, $m \geq 1$.

We will use type b) XIP-neofields of orders $v = 6m+2$ (with the exception whem $m = 1$; in that case we only have type a) XIP-neofields of order 8) to construct type b) XIP-neofields of orders $v = 18m+4$. The existence of type b) XIP-neofields of orders $v = 6m+2$ is guaranteed by the construction in Case 1.

In this recurrence construction, we have $v = 18m+4$, $m \geq 2$. We partition $\mathbf{Z}^*_{18m+3}$ into $6m$ sextuples with 3 sextuples derived from each of the $2m$ sextuples in any type b) XIP-admissible partition of $\mathbf{Z}^*_{6m+1}$. Since 3 divides $v-1 = 18m+3$, let

$$H = \left\{0, \frac{v-1}{3}, \frac{2(v-1)}{3}\right\} = \{0, 6m+1, 12m+2\} \cong \mathbf{Z}_3.$$

By taking the quotient of $\mathbf{Z}_{18m+3}$ and H, we have the cosets

$$\mathbf{Z}_{18m+3}/H = \{H, H+1, H+2, \ldots, H+6m\} \cong \mathbf{Z}_{6m+1}.$$

Let $\Pi = \{S(k_1,n_1), S(k_2,n_2), \ldots, S(k_{2m},n_{2m})\}$ be any type b) XIP-admissible partition of $\mathbf{Z}^*_{6m+1}$. For each $S(k,n) \in \Pi$, letting a = 6m+1, we have

$$H+k = \{k,\ a+k,\ 2a+k\}$$
$$H+n = \{n,\ a+n,\ 2a+n\}$$

and we define T(k,n) and R(k,n) as

$$T(k,n) = \{a+k,\ 2a+n,\ a+(n-k),\ 2a+(-k),\ a+(-n),\ 2a+(k-n)\}$$

and

$$R(k,n) = \{2a+k,\ a+n,\ 2a+n-k),\ a+(-k),\ 2a+(-n),\ a+(k-n)\}.$$

For every S(k,n) in π, let S*(k,n) be the sextuple generated by (k,n) in $\mathbf{Z}_{18m+3}$. We claim that

$$\begin{aligned}\pi^* = &\{S^*(k_1,n_1),\ \ldots,\ S^*(k_{2m},n_{2m})\}\\ &\cup\{T(k_1,n_1),\ \ldots,\ T(k_{2m},n_{2m})\}\\ &\cup\{R(k_1,n_1),\ \ldots,\ R(k_{2m},n_{2m})\}\\ &\cup\{a,2a\}\end{aligned}$$

is a type b) XIP-admissible partition of $\mathbf{Z}^*_{18m+3}$.

We have the following "nesting" procedure

$$\begin{cases}S^*(k_1,n_1)\\ T(k_1,n_1)\\ R(k_1,n_1)\end{cases} \qquad \begin{cases}S^*(k_{m+1},n_{m+1})\\ T(k_{m+1},n_{m+1})\\ R(k_{m+1},n_{m+1})\end{cases}$$

$$\begin{cases}S^*(k_2.n_2)\\ T(k_2,n_2)\\ R(k_2,n_2)\end{cases} \qquad \begin{cases}S^*(k_{m+2},n_{m+2})\\ T(k_{m+2},n_{m+2})\\ R(k_{m+2},n_{m+2})\end{cases}$$

$$\vdots \qquad\qquad \vdots$$

$$\begin{cases} S^*(k_m,n_m) \\ T(k_m,n_m) \\ R(k_m,n_m) \end{cases} \qquad \begin{cases} S^*(k_{2m},n_{2m}) \\ T(k_{2m},n_{2m}) \\ R(k_{2m},n_{2m}) \end{cases}$$

$\{a,2a\}$.

It is clear that sextuples $S^*(k,n)$, $T(k,n)$ and $R(k,n)$ are disjoint. Moreover, there don't exist any two equal sextuples in the above nesting.

Let x be any element in $\mathbb{Z}_{18m+3}$-H. Then x must be in some coset, say in $H+k = \{k, a+k, 2a+k\}$. Hence there exist sextuples $S(k,n)$ and $S(n',k)$ in π. If $x = k$, then x is of odd-parity in $S^*(k,n)$ and is of even-parity in $S^*(n',k)$ in π^*. If $x = a+k$, then x is of odd-parity in $T(k,n)$ and is of even-parity in $R(n',k)$ in the partition π^*. If $x = 2a+k$, then x is an odd-parity element in $R(k,n)$ and is an even-parity element in $T(n',k)$ in the partition π^*. Hence the partition π^* satisfies b) of Definition I.43, i.e., π^* is a type b) XIP-admissible partition of $\mathbb{Z}^*_{18m+3}$. By Definition I.45 and Theorem I.46, π^* determines a type b) XIP-neofields of order $v = 18m+4$, $m \geq 2$.

We give two examples to illustrate the above construction. Note that the construction fails to work for the case $m = 1$. Therefore we present a type b) XIP-neofield of order 22 by "trial and error" construction.

Examples:

For $v = 22$, we have the following type b)

XIP-admissible partition of $\mathbb{Z}_{21}^{*}$.

$$S(2,10) = \{2, 10, 8, 19, 11, 13\}$$
$$S(4,9) = \{4, 9, 5, 17, 12, 16\}$$
$$S(6,1) = \{6, 1, 16, 15, 20, 5\}$$
$$S(9,3) = \{9, 3, 15, 12, 18, 6\}$$
$$S(1,4) = \{1, 4, 3, 20, 17, 18\}$$
$$S(10,2) = \{10, 2, 13, 11, 19, 8\}$$
$$S(7,14) = \{7, 14\}.$$

Hence by Definition I.45 b) and Theorem I.46, we have the presentation function for a type b) XIP-neofield N_{22}.

x	0	1	a	a^2	a^3	a^4	a^5	a^6	a^7	a^8	a^9	a^{10}	a^{11}	a^{12}	a^{13}
$T(x)$	1	0	a^4	a^{10}	a^{20}	a^9	a^{17}	a	a^{14}	a^{19}	a^3	a^2	a^{13}	a^{16}	a^{11}

x	a^{14}	a^{15}	a^{16}	a^{17}	a^{18}	a^{19}	a^{20}
$T(x)$	a^7	a^{17}	a^{15}	a^{18}	a^6	a^8	a^5

For $v = 40$, $m = 2$, 4 sextuples for a type b) XIP-admissible partition π of $\mathbb{Z}_{13}^{*}$ are required.

$S(1,4) = \{1, 4, 3, 12, 9, 10\}$ $S(5,2) = \{5, 2, 10, 8, 11, 3\}$

$S(2,6) = \{2, 6, 4, 11, 7, 9\}$ $S(6,1) = \{6, 1, 8, 7, 12, 5\}$

By taking $H = \{0, 13, 26\} \cong \mathbb{Z}_3$, we have

$$\mathbb{Z}_{39}/H = \{H, H+1, \ldots, H+12\} \cong \mathbb{Z}_{13},$$

where $H+k = \{k, 13+k, 26+k\}$.

By deriving S*(k,n), T(k,n) and R(k,n) in $\mathbf{Z}_{39}$ from S(k,n) of π in $\mathbf{Z}_{13}$, we have a type b) XIP-admissible partition of $\mathbf{Z}_{39}$:

$$\begin{cases} S^*(1,4) = \{1, 4, 3, 38, 35, 36\} \\ T(14,30) = \{14, 30, 16, 25, 9, 23\} \\ R(27,17) = \{27, 17, 29, 12, 22, 10\} \end{cases}$$

$$\begin{cases} S^*(5,2) = \{5, 2, 36, 34, 37, 3\} \\ T(18,28) = \{18, 28, 10, 21, 11, 29\} \\ R(31,15) = \{31, 15, 23, 8, 24, 16\} \end{cases}$$

$$\begin{cases} S^*(2,6) = \{2, 6, 4, 37, 33, 35\} \\ T(15,32) = \{15, 32, 17, 24, 7, 22\} \\ R(28,19) = \{28, 19, 30, 11, 20, 9\} \end{cases}$$

$$\begin{cases} S^*(6,1) = \{6, 1, 34, 33, 38, 5\} \\ T(19,27) = \{19, 27, 8, 20, 12, 31\} \\ R(32,14) = \{32, 14, 21, 7, 25, 18\}. \end{cases}$$

$$S^*(13,26) = \{13,26\}.$$

For v = 58, m = 3, 6 sextuples for a type b) XIP-admissible partition π of $\mathbf{Z}^*_{19}$ are required.

$$S(1,5) = \{1, 5, 4, 18, 14, 15\}$$

$$S(2,7) = \{2, 7, 5, 17, 12, 14\}$$

$$S(3,9) = \{3, 9, 6, 16, 10, 13\}$$

$$S(9,1) = \{9, 1, 11, 10, 18, 8\}$$

$$S(8,2) = \{8, 2, 13, 11, 17, 6\}$$

$$S(7,3) = \{7, 3, 15, 12, 16, 4\}$$

By taking $H = \{0, 19, 38\} \cong \mathbb{Z}_3$, we have

$$\mathbb{Z}_{57}/_{H} = \{H, H+1, \ldots, H+18\} \cong \mathbb{Z}_{19},$$

where $H+k = \{k, 19+k, 38+k\}$. After constructing $S^*(k,n)$, $T(k,n)$ and $R(k,n)$ in $\mathbb{Z}_{57}$ from $S(k,n)$ of π in $\mathbb{Z}_{19}$, we have a type b) XIP-admissible partition of $\mathbb{Z}_{57}$:

$$\begin{cases} S^*(1,5) = \{1, 5, 4, 56, 52, 53\} \\ T(20,43) = \{20, 43, 23, 37, 14, 34\} \\ R(39,24) = \{39, 24, 42, 18, 33, 15\} \end{cases}$$

$$\begin{cases} S^*(2,7) = \{2, 7, 5, 55, 50, 52\} \\ T(21,45) = \{21, 45, 24, 36, 12, 33\} \\ R(40,26) = \{40, 26, 43, 17, 31, 14\} \end{cases}$$

$$\begin{cases} S^*(3,9) = \{3, 9, 6, 54, 48, 51\} \\ T(22,47) = \{22, 47, 25, 35, 10, 32\} \\ R(41,28) = \{41, 28, 44, 16, 29, 13\} \end{cases}$$

$$\begin{cases} S^*(9,1) = \{9, 1, 49, 48, 56, 8\} \\ T(28,39) = \{28, 39, 11, 29, 18, 46\} \\ R(47,20) = \{47, 20, 30, 10, 37, 27\} \end{cases}$$

$$\begin{cases} S^*(8,2) = \{8, 2, 51, 49, 55, 6\} \\ T(27,40) = \{27, 40, 13, 30, 17, 44\} \\ R(46,21) = \{46, 21, 32, 11, 36, 25\} \end{cases}$$

$$\begin{cases} S^*(7,3) = \{7, 3, 53, 50, 54, 4\} \\ T(26,41) = \{26, 41, 15, 31, 16, 42\} \\ R(45,22) = \{45, 22, 34, 12, 35, 23\}. \end{cases}$$

$$S^*(19,38) = \{19, 38\}$$

The above constructions provide the result below.

<u>Theorem II.1</u>. There exists a type b) XIP-neofield N_v of even order v if $v \equiv 2, 4, 8, 14 \pmod{18}$.

Sec. 2. <u>Constructions of Odd Order Type b) XIP-Neofields</u>

In this section, constructions and examples are provided verifying the existence of odd order type b) (in the sense of Definition I.55) XIP-neofields of orders $v \equiv 3, 5 \pmod 6$ and $v \equiv 7 \pmod{18}$.

<u>Case 1</u>: $v = 6m+3$, $m \geq 2$, $\ell = 3m+1$.

We partition $\mathbb{Z}^*_{6m+2}$ into 2m sextuples $\{k, n, n-k, -k, -n, k-n\}$ which give a type b) XIP-admissible partition of $\mathbb{Z}^*_{6m+2}$.

First, pair k and n by scheme OA1:

k	n	n-k	-k	-n	k-n
1	m+2	m+1	6m+1	5m	5m+1
2	m+4	m+2	6m	5m=2	5m
3	m+6	m+3	6m=1	5m-4	5m-1
.	.	.	.	.	.
.	.	.	.	.	.
.	.	.	.	.	.
m-1	3m-2	2m-1	5m+3	3m+4	4m+3
m	3m	2m	5m+2	3m+2	4m+2

<u>OA1</u>

yielding by a "nesting" procedure odd-parity elements

$$\left\{k,\ n-k,\ -n\right\}$$

for each odd element (even element) from 3m+2 to 5m when m is odd (when m is even) and all elements from 1 to 2m; and even-parity elements

$$\left\{n,\ -k,\ k-n\right\}$$

for each odd element (even element) from m+2 to 3m when m is odd (when m is even) and elements from 4m+2 to 6m+1.

Then we pick k and n by scheme OB1:

k	n	n-k	-k	-n	k-n
3m	1	3m+3	3m+2	6m+1	3m-1
3m-1	2	3m+5	3m+3	6m	3m-3
3m-2	3	3m+7	3m+4	6m-1	3m-5
.	.	.	.	.	.
.	.	.	.	.	.
.	.	.	.	.	.
2m+2	m-1	5m-1	4m	5m+3	m+3
2m+1	m	5m+1	4m+1	5m+2	m+1

OB1

yielding by a "nesting" procedure odd-parity elements

$$\left\{k,\ n-k,\ -n\right\}$$

for each even element (odd element) from 3m+3 to 5m+1 when m is odd (when m is even) and all elements from 2M+1 to 3m and from 5m+2 to 6m+1; and even-parity elements

$$\{k,\ n-k,\ -n\}$$

for each even element (odd element) from 3m+3 to 5m+1 when m is odd (when m is even) and all elements from 2m+1 to 3m and from 5m+2 to 6m+1; and even-parity elements

$$\{n,\ -k,\ k-n\}$$

for each even element (odd element) from m+1 to 3m-1 when m is odd (when m is even) and all elements from 1 to m and from 3m+2 to 4m+1.

One checks that the set of all the odd-parity elements (or all the even-parity elements) is equal to $\mathbb{Z}^*_{6m+12}$. Moreover, one checks that in this construction, $|S(1,m+2) \cap S(3m,1)| = 2$, so the partition is not of type a) in the sense of Definition I.55. Hence we have a type b) XIP-admissible partition of $\mathbb{Z}^*_{6m+1}$, and by Definition I.58 b) and Theorem I.59, this determines a type b) XIP-neofield of order v = 6m+3, m ≥ 2.

This construction requires that m ≥ 2: For m = 1, we have v = 9. There are exactly four XIP-neofields of order 9, two of them have CIP and the other two have SIP. Hence there doesn't exist any type b) XIP-neofield of order 9.

Examples:

For m = 2, v = 15 we have 4 sextuples and one singleton.

$S(1,4) = \{1, 4, 3, 13, 10, 11\}$ $S(6,1) = \{6, 1, 9, 8, 13, 5\}$
$S(2,6) = \{2, 6, 4, 12, 8, 10\}$ $S(5,2) = \{5, 2, 11, 9, 12, 3\}$
$S(0,0) = \{0\}$

in a type b) XIP-admissible partion of $\mathbf{Z}_{14}^*$.

For m = 3, v = 21 we have 6 sextuples and one singleton.

$S(1,5) = \{1, 5, 4, 19, 15, 16\}$, $S(9,1) = \{9, 1, 12, 11, 19, 8\}$
$S(2,7) = \{2, 7, 5, 18, 13, 15\}$, $S(8,2) = \{8, 2, 14, 12, 18, 6\}$
$S(3,9) = \{3, 9, 6, 17, 11, 14\}$, $S(7,3) = \{7, 3, 16, 13, 17, 4\}$
$S(0,0) = \{0\}$

in a type b) XIP-admissible partition of $\mathbf{Z}_{20}^*$.

For m = 4, v = 27, we have 8 sextuples and one singleton

$$S(1,6) = \{1, 6, 5, 25, 20, 21\}$$
$$S(2,8) = \{2, 8, 6, 24, 18, 20\}$$
$$S(3,10) = \{3, 10, 7, 23, 16, 19\}$$
$$S(4,12) = \{4, 12, 8, 22, 14, 18\}$$
$$S(12,1) = \{12, 1, 15, 14, 25, 11\}$$
$$S(11,2) = \{11, 2, 17, 15, 24, 9\}$$
$$S(10,3) = \{10, 3, 19, 16, 23, 7\}$$
$$S(9,4) = \{9, 4, 21, 17, 22, 5\}$$
$$S(0,0) = \{0\}.$$

in a type b) XIP-admissible partition of $\mathbf{Z}_{26}^*$.

Note that the pattern of the construction is

somewhat similar to that of the case v = 6m+2 in Section 1 of this chapter. Hence when v = 6m+3 and m = 3f+1, we have

$$S(s,t) = \{2f+1,\ 7f+3,\ 5f+2,\ 16f+8,\ 11f+5,\ 13f+7\}$$

in scheme OA1 and S(t,s) in scheme OB1. It follows that $t \geq 1$ in the sense of Definition I.55.

We now turn to the construction of the case when $v \equiv 5 \pmod 6$.

Case 2: $v = 6m+5$, $m \geq 2$, $\ell = 3m+2$

We partition $\mathbb{Z}^*_{6m+4}$ into 2m sextuples and one triple, yielding a type b) XIP-admissible partition of $\mathbb{Z}^*_{6m+4}$.

First, pair k and n by scheme OA2:

k	n	n-k	-k	-n	k-n
1	m+2	m+1	6m+3	5m+2	5m+3
2	m+4	m+2	6m+2	5m	5m+2
3	m+6	m+3	6m+1	5m-2	5m+1
.	.	.	.	.	.
.	.	.	.	.	.
.	.	.	.	.	.
m-1	3m-2	2m-1	5m+5	3m+6	4m+5
m	3m	2m	5m+4	3m+4	4m+4

OA2

and we pair k and n by the scheme OB2:

k	n	n-k	-k	-n	k-n
3m	1	3m+5	3m+4	6m+3	3m-1
3m-1	2	3m+7	3m+5	6m+2	3m-3
3m-2	3	3m+0	3m+6	6m+1	3m-5
.	.	.	.	.	.
.	.	.	.	.	.
.	.	.	.	.	.
2m+2	m-1	5m+1	4m+2	5m+5	3m-5
2m+1	m	5m+3	4m+3	5m+4	m+1

OB2

Combining schemes OA2 and OB2 and the triple:

$$S(0,3m+1) = \{0,\ 3m+1,\ 3m+3\},$$

We have a type b) XIP-admissible partition by the same argument as we have in Case 1. Hence by Definition I.58 b) and Theorem I.59, this construction determines a type b) XIP-neofield of order $v = 6m+5$, $m \geq 2$.

Example: We only give one example to illustrate the above construction. Note that the above construction gives rise to a type a) XIP-neofield of order 11 when m = 1.

For m = 3, v = 23 and ℓ = 11; we have the following 6 sextuples and a triple:

$S(1,5) = \{1, 5, 4, 21, 17, 18\}$, $S(9,1) = \{9, 1, 14, 13, 21, 8\}$
$S(2,7) = \{2, 7, 5, 20, 15, 17\}$, $S(8,2) = \{8, 2, 16, 14, 20, 6\}$
$S(3,9) = \{3, 9, 6, 19, 13, 16\}$, $S(7,3) = \{7, 3, 18, 15, 19, 4\}$

$S(0,10) = \{0, 10, 12\}$,

these determine a type b) XIP-neofield of order 23.

Case 3: $v = 18m+7$, $m \geq 2$.

Similar to Case 2 in Section III.1 (except that we have a special triple and a pair in this case), we use a type b) XIP-neofield of order $v = 6m+3$ to construct type b) XIP-neofield of order $v = 18m+7$. The existence of type b) XIP-neofields of orders $v = 6m+3$ are guaranteed by Case 1 of this section. Note that we need a type b) XIP-neofield of order 9 to construct a type b) XIP-neofield of order 25. Hence this method doesn't apply in the case $m = 1$.

We give one example to illustrate this construction.

Example: For $m = 2$, $v = 43$ and $\ell = 21$, 4 sextuples and a singleton constitute a type b) XIP-admissible partition of $\mathbb{Z}_{14}^{*}$.

$S(1,4) = \{1, 4, 3, 13, 10, 11\}$, $S(6,1) = \{6, 1, 9, 8, 13, 5\}$

$S(2,6) = \{2, 6, 4, 12, 8, 10\}$, $S(5,2) = \{5, 2, 11, 9, 12, 3\}$

$S(0,0) = \{0\}$.

By taking $H = \{0, 14, 28\} \cong \mathbb{Z}_3$, we have

$$\mathbb{Z}_{42}/H = \{H, H+1, \ldots, H+13\} \cong \mathbb{Z}_{14},$$

where $H+k = \{k, 14+k, 28+k\}$. We then have a type b) XIP-admissible partition of $\mathbb{Z}_{42}^{*}$,

$$\begin{cases} S^*(1,4) = \{1, 4, 3, 41, 38, 39\} \\ T(15,32) = \{15, 32, 17, 27, 10, 25\} \\ R(29,18) = \{29, 18, 31, 13, 24, 11\} \end{cases}$$

$$\begin{cases} S^*(2,6) = \{2, 6, 4, 40, 36, 38\} \\ T(16,34) = \{16, 34, 18, 26, 8, 24\} \\ R(30,20) = \{30, 20, 32, 12, 22, 10\} \end{cases}$$

$$\begin{cases} S^*(6,1) = \{6, 1, 37, 36, 41, 5\} \\ T(20,24) = \{20, 29, 9, 22, 13, 33\} \\ R(34,15) = \{34, 15, 23, 8, 27, 19\} \end{cases}$$

$$\begin{cases} S^*(5,2) = \{5, 2, 39, 37, 40, 3\} \\ T(19,30) = \{19, 30, 11, 23, 12, 31\} \\ R(33,16) = \{33, 16, 25, 9, 26, 17\} \end{cases}$$

$$S^*(0,7) = \{0, 7, 35\}$$

$$S^*(14,28) = \{14, 28\}.$$

By Definition I.58 b) and Theorem I.59, this determines a type b) XIP-neofield of order 43.

Taken together, the three cases so far elaborated imply the following result.

<u>Theorem III.2</u>: There exists a type b) XIP-neofield N_v of odd order v if $v \equiv 3, 5, 7, 9, 11, 15, 17 \pmod{18}$.

All the cases considered in both sections of this chapter yield the following result:

<u>Theorem III.3</u>: For $v \equiv 2, 3, 4, 5, 7, 8, 9, 11, 14, 15, 17 \pmod{18}$, there exists a type b) XIP-neofield of order v.

Hence by Theorem I.41, Definition II.19, Example II.26, Remark II.27 and the above theorem, we have the following main results:

Theorem III.4: There does not exist a pure proper XIP neofield of any order $v \equiv 0 \pmod 6$ or $v \leq 10$. For any other order v, there exists a pure proper XIP-neofield, N_v.

Theorem III.5: There does not exist a proper XIP neofield of any order $v \equiv 0 \pmod 6$, or $v \leq 7$, or $v = 10$. For any other order v, there exists a proper XIP-neofield, N_v.

CHAPTER IV

CONSTRUCTION OF PROPER LXP- AND PROPER XMP-NEOFIELDS

Sec. 1. Combinatorial Construction of Proper LXP-Neofields

In this section, combinatorial constructions and examples are provided verifying the existence of proper LXP-neofields (a proper LXP-neofield is an LXP-neofield which doesn't have CIP) of all orders $v \geq 6$. Note that we have LXP-neofield of orders $v = 4$ and $v = 5$. But they all have CIP (since they are fields).

Case 1: $v = 4m$, $m \geq 3$.

We partition $\mathbb{Z}^*_{4m-1}$ into 2m-1 unordered pairs $\{k,n\}$ which give an LXP-admissible partition of $\mathbb{Z}^*_{4m-1}$.

We pick k and n by the following scheme LA1:

k	n	n-k
2	4m-2	4m-4
3	4m-3	4m-6
.	.	.
.	.	.
.	.	.
m	3m	2m
m+1	3m-3	2m-4
m+2	3m-4	2m-6

(continued)

k	n	n-k
.	.	.
.	.	.
.	.	.
2m-2	2m	2

LA1

yielding by a "nesting" procedure of the elements

$$\{\pm(n-k)\}$$

for each odd element from 3 to 4m-3 (except 2m+1) and for each even element from 2 to 4m-4 (except 2m-2). Then we have the two triples

k	n	n-k
1	2m-1	2m-2
3m-2	3m-1	1

which give the elements $\{\pm(n-k)\}$ for 1, 2m-2, 2m+1, 4m-2.

One checks that the differences $\{\pm(n-k)\}$ range over all non-zero elements in $\mathbf{Z}_{4m-1}$. Moreover, these 2m-1 unordered pairs $\{k,n\}$ are disjoint. By Definition I.29, this gives an LXP-admissible partition of $\mathbf{Z}^*_{4m-1}$. Hence by Definition I.31 and Theorem I.32, this construction determines an LXP-neofield of order v = 4m, m ≥ 3. One can easily check that these LXP-neofields are proper.

Examples:

For m = 3, v = 12, we have 5 unordered pairs

$\{k,n\}$:

k	n	n-k
2	10	8
3	9	6
4	6	2
- - -	- - -	- - -
1	5	4
7	8	1.

Hence we have a proper LXP-neofield of order 12 with the following presentation function:

x	0	1	a	a^2	a^3	a^4	a^5	a^6	a^7	a^8	a^9	a^{10}
T(x)	1	0	a^5	a^{10}	a^9	a^6	a	a^4	a^8	a^7	a^3	a^2

.

For m = 4, v = 16, we have 7 unordered pairs $\{k,n\}$:

k	n	n-k
2	14	12
3	13	10
4	12	8
5	9	4
6	8	2
- - -	- - -	- - -
1	7	6
10	11	1.

Hence we have a proper LXP-neofield of order 16 with the following presentation function:

x	0	1	a	a^2	a^3	a^4	a^5	a^6	a^7	a^8	a^9	a^{10}	a^{11}	a^{12}	a^{13}	a^{14}
$T(x)$	1	0	a^7	a^{14}	a^{13}	a^{12}	a^9	a^8	a	a^6	a^5	a^{11}	a^{10}	a^4	a^3	a^2 .

Not covered by the above construction is the case $v = 8$, for which we supply the example here with the presentation function:

x	0	1	a	a^2	a^3	a^4	a^5	a^6
$T(x)$	1	0	a^3	a^6	a	a^5	a^4	a^2 .

Case 2: $v = 4m+2$, $m \geq 3$.

In this case we obtain a partition of $\mathbb{Z}^*_{4m+1}$ into $2m$ unordered pairs $\{k,n\}$ which gives an LXP-admissible partition of $\mathbb{Z}^*_{4m+1}$.

We pick k and n by the following scheme LA2:

k	n	n-k
2	4m	4m-2
3	4m-1	4m-4
4	4m-2	4m-6
.	.	.
.	.	.
.	.	.
m-1	3m+3	2m+4
m	3m	2m
m+1	3m-1	2m-2
m+2	3m-2	2m-4

(continued)

$\underline{k}$	$\underline{n}$	$\underline{n-k}$
.	.	.
.	.	.
.	.	.
2m-1	2m+1	2

LA2

yielding by a "nesting" procedure of the elements

$$\{\pm(n-k)\}$$

for each odd element from 3 to 4m-1 (except 2m-1) and for each even element from 2 to 4m-2 (except 2m+2). Then we have the two triples:

$\underline{k}$	$\underline{n}$	$\underline{n-k}$
1	2m	2m-1
3m+1	3m+2	1 ,

which give the elements $\{\pm(n-k)\}$ for 1, 2m-1, 2m+2, 4m.

Once checks that the differences $\{\pm(n-k)\}$ range over all non-zero elements in $\mathbf{Z}_{4m-1}$ and these 2m unordered pairs $\{k,n\}$ are disjoint. By Definition I.29, Definition I.31 and Theorem I.32, this gives rise to a proper LXP-neofield of order v = 4m+2, m $\geq$ 3.

Examples:

For m = 3, v = 14, we have 6 unordered pairs $\{k,n\}$:

k	n	n-k
2	12	10
3	4	6
4	8	4
5	7	2
- - -	- - -	- - -
1	6	5
10	11	1

Hence we have a proper LXP-neofield of order 14 with the following presentation function:

x	0	1	a	a^2	a^3	a^4	a^5	a^6	a^7	a^8	a^9	a^{10}	a^{11}	a^{12}
$T(x)$	1	0	a^6	a^{12}	a^9	a^8	a^7	a	a^5	a^4	a^3	a^{11}	a^{10}	a^2

For $m = 5$, $v = 22$, we have 10 unordered pairs $\{k,n\}$:

k	n	n-k
2	20	18
3	19	16
4	18	14
5	15	10
6	14	8
7	13	6
8	12	4
9	11	2
- - -	- - -	- - -
1	10	9
16	17	1

We then have the presentation function of a proper LXP-neofield of order 22.

x	0	1	a	a^2	a^3	a^4	a^5	a^6	a^7	a^8	a^9	a^{10}	a^{11}	a^{12}
$T(x)$	1	0	a^{10}	a^{20}	a^{19}	a^{18}	a^{15}	a^{14}	a^{13}	a^{12}	a^{11}	a	a^9	a^8

x	a^{13}	a^{14}	a^{15}	a^{16}	a^{17}	a^{18}	a^{19}	a^{20}
$T(x)$	a^7	a^6	a^5	a^{17}	a^{16}	a^4	a^3	a^2

The construction fails for $m = 1$ and $m = 2$, so we supply these two examples of proper LXP-neofields.

$v = 6$,

x	0	1	a	a^2	a^3	a^4
$T(x)$	1	0	a^4	a^3	a^2	a

$v = 10$,

x	0	1	a	a^2	a^3	a^4	a^5	a^6	a^7	a^8
$T(x)$	1	0	a^4	a^6	a^5	a	a^3	a^2	a^8	a^7

Note that by Theorem III.5, there doesn't exist a proper XIP-neofield of order 10. Also from [4], there doesn't exist a CIP-neofield of order 10. But proper LXP-neofields do exist as shown above. In fact, there are 9 proper LXP-neofields (by Computer Search).

We now turn our attention to the case of odd order.

Case 3: $v = 4m+1$, $m \geq 2$.

We partition $\mathbb{Z}_{4m}^*$ into 2m-1 unordered pairs $\{k,n\}$ and a singleton $\{m\}$ which give an LXP-admissible partition of $\mathbb{Z}_{4m}^*$.

We pick k and n by the following scheme LB1.

$\underline{k}$	$\underline{n}$	$\underline{n-k}$
0	4m-1	4m-1
1	4m-2	4m-3
2	4m-3	4m-5
.	.	.
.	.	.
.	.	.
m-1	3m	2m+1
m+1	3m-1	2m-4
.	.	.
.	.	.
.	.	.
2m-1	2m+1	2
- - - -	- - - -	- - - -
m	m	0

LB1

yielding by a "nesting" procedure of the elements

$$\left\{ \pm(n-k) \right\}$$

for each element from 1 to 4m-1 (except 2m) and the element 0 in $\mathbb{Z}_{4m}$. Moreover, these 2m-1 unordered pairs $\{k,n\}$ are disjoint. By Definition I.29, this gives an LXP-admissible partition of $\mathbb{Z}_{4m}$. Hence by Definition I.31 and Theorem I.32, this construction determines an LXP-neofield of odd order v = 4m+1, m $\geq$ 2. One can easily check that these LXP-neofields are proper.

Examples:

For m = 2, v = 9, we have 3 unordered pairs $\{k,n\}$ and a singleton $\{2\}$.

k	n	n-k
0	7	7
1	6	5
3	5	2
- - -	- - -	- - -
2	2	0

Hence we have a proper LXP-neofield, N_9.

x	0	1	a	a^2	a^3	a^4	a^5	a^6	a^7
$T(x)$	1	a^3	a^2	a^6	a	0	a^7	a^5	a^4

For m = 4, v = 17, we have 7 unordered pairs $\{k,n\}$ and a singleton $\{4\}$.

k	n	n-k
0	15	15
1	14	13
2	13	11
3	12	9
5	11	6
6	10	4
7	9	2
- - -	- - -	- - -
4	4	0

We then have the presentation function of a proper LXP-neofield, N_{17}.

x	0	1	a	a^2	a^3	a^4	a^5	a^6	a^7	a^8	a^9	a^{10}	a^{11}	a^{12}	a^{13}
$T(x)$	1	a^7	a^6	a^5	a^4	a^{12}	a^3	a^2	a	0	a^{15}	a^{14}	a^{13}	a^{11}	a^{10}

x	a^{14}	a^{15}
$T(x)$	a^9	a^8 .

Case 4: $v = 4m$, $m \geq 2$.

We partition $\mathbb{Z}^*_{4m+2}$ into $2m$ unordered pairs $\{k,n\}$ and a singleton $\{m\}$ which give an LXP-admissible partition of $\mathbb{Z}^*_{4m+2}$.

We pick k and n by the following scheme LB2:

k	n	n-k
0	4m+1	4m+1
1	4m	4m-1
2	4m-1	4m-3
.	.	.
.	.	.
.	.	.
m-1	3m+2	2m+3
m+1	3m+1	2m
m+2	3m	2m-2
.	.	.
.	.	.
.	.	.
2m	2m+2	2

\- - - - - - - - - - (continued)

$\underline{k}$	$\underline{n}$	$\underline{n-k}$
m	m	0

LB2

yielding by a "nesting" procedure of the elements

$$\{\pm(n-k)\}$$

for each element from 1 to 4m+1 (except 2m+1) and the element 0 in $\mathbf{Z}_{4m+2}$. Moreover, these 2m unordered pairs $\{k,n\}$ are disjoint. By Definition I.29, Definition I.31, and Theorem I.32, this construction determines a proper LXP-neofield of odd order v = 4m+3, m ≥ 2.

Examples:

For m = 2, v = 11, we have 4 unordered pairs $\{k,n\}$ and a singleton $\{2\}$.

$\underline{k}$	$\underline{n}$	$\underline{n-k}$
0	9	9
1	8	7
3	7	4
4	6	2
- - -	- - -	- -
2	2	0 .

We list the presentation function of this proper LXP-neofield, N_{11}.

x	0	1	a	a^2	a^3	a^4	a^5	a^6	a^7	a^8	a^9
T(x)	1	a^4	a^3	a^7	a^9	a	0	a^9	a^8	a^6	a^5.

For $m = 3$, $v = 15$, we have 6 unordered pairs $\{k,n\}$ and the singleton $\{3\}$.

k	n	$n-k$
0	13	13
1	12	11
2	11	9
4	10	6
5	9	4
6	8	2
- - -	- - -	- - -
3	3	0

Hence we have the presentation function of a proper LXP-neofield, N_{15}.

x	0	1	a	a^2	a^3	a^4	a^5	a^6	a^7	a^8	a^9	a^{10}	a^{11}	a^{12}	a^{13}
$T(x)$	1	a^6	a^5	a^4	a^{10}	a^3	a^2	a	0	a^{13}	a^{12}	a^{11}	a^9	a^8	a^7 .

The construction fails for $m = 1$, i.e., for a proper LXP-neofield N_7. We supply an example obtained by trial.

x	0	1	a	a^2	a^3	a^4	a^5
$T(x)$	1	a^2	a^4	a	0	a^5	a^3

All the cases considered in this section yield the following result:

<u>Theorem IV.1</u>: For any integer $v \geq 6$, there exists a proper LXP-neofield of order v.

Sec. 2. Construction of Even Order LXP-Neofields by a Number Theory Technique

In this section, we apply number theory (see [1]) to give a construction of even order LXP-neofields N_v with $p = v-1 = 2^m \cdot q+1$, a prime and q an odd number. For the case p = v-1 not a prime, we have the product theorem in the next section and examples are given.

Let $p = 2^m \cdot q+1$, q odd and $q > 1$. Let x be a primitive root of the Galois field $\mathbb{Z}_p$, where $p = 2^m \cdot q+1$ and q is odd. Let us consider the following $2^{m-1} \cdot q$ pairs $\{k,n\}$.

$$\pi(s,t) = \left\{x^{(2s-2)2^{m-1}+t},\ x^{(2s-1)2^{m-1}+t}\right\}$$

where s = 1, 2, 3, ..., q and t = 1, 2, ..., 2^{m-1}.

First of all, we prove that for a fixed s or a fixed t, $\pi(s,t)$ are distint. Assume

$$(2s-2)2^{m-1}+t_1 \equiv (2s-1)2^{m-1}+t_2 \pmod p$$

Then $(t_1-t_2) \equiv 2^{m-1} \pmod p$, which is impossible since $t_1-t_2 < 2^{m-1}$. Also we assume

$$(2s_1-2)2^{m-1}+t \equiv (2s_2-1)2^{m-1}+t \pmod p,$$

and without loss of generality we assume $s_1 > s_2$. Then we have

$$2(s_1-s_2) \equiv 1 \pmod p.$$

This contradicts $2(s_1-s_2) < 2q$.

Next, we claim that all the differences $\{\pm(n-k)\}$ are distinct. For if not, we have

$$x^{(2s_1-2)2^{m-1}+t_1} - x^{(2s_1-1)2^{m-1}+t_1} \equiv \pm (x^{(2s_2-2)2^{m-1}+t_2} - x^{(2s_2-1)2^{m-1}+t_2}) \pmod p$$

Again we assume $s_1 > s_2$. If $t_1 > t_2$, we have:

$$(x^{(2s_1-2s_2)2^{m-1}+(t_1-t_2)} \pm 1)(1-x^{2^{m-1}}) \cdot x^{(2s_2-2)2^{m-1}+t_2} \equiv 0.$$

If $t_1 < t_2$, we have:

$$(x^{(2s_1-2s_2)2^{m-1}-(t_2-t_1)} \pm 1)(1-x^{2^{m-1}}) \cdot x^{(2s_2-2)2^{m-1}+t_2} \equiv 0.$$

Hence we have the following possibilities:

a) $1-x^{2^{m-1}} \equiv 0 \pmod p$; this is false because

$x^{2^m \cdot q} \equiv 1 \pmod p$ (by Fermat's Little Theorem).

b) $x^{2(s_2-2)\cdot 2^{m-1}+t_2} \equiv 0 \pmod p$; this would imply

$x = 0$.

c) $x^{(2s_1-2s_2)2^{m-1}+(t_1-t_2)} \equiv \pm 1 \pmod p$; since

$2(s_1-s_2) < 2q$ and $t_1-t_2 < 2^{m-1}$, we have

$(2s_1-2s_2)2^{m-1}+(t_1-t_2) < 2q\cdot 2^{m-1} + 2^{m-1}$

$< 2^m\cdot q + 2^{m-1}\cdot q$

If the congruence equation holds, then we have

$$(2s_1-2s_2)2^{m-1}+(t_1-t_2) = 2^m q \text{ or } 2^{m-1}q.$$

Hence

$$\begin{aligned}(t_1-t_2) &= 2^m q-(2s_1-2s_2)2^{m-1} \text{ or } 2^{m-1}q-(2s_1-2s_2)2^{m-1} \\ &= 2^m[q-(s_1-s_2)] \text{ or } 2^{m-1}[q-2(s_1-s_2)],\end{aligned}$$

i.e., $t_1-t_2 > 2^m$ or 2^{m-1} (since q is odd). In either case, it leads to a contradiction.

d) $x^{(2s_1-2s_2)2^{m-1}-(t_2-t_1)} \equiv \pm 1 \pmod p$; by the same reason as in c), we have:

$$t_2-t_1 = 2^m[(s_1-s_2)-q] \text{ or } 2^{m-1}[2(s_1-s_2)-q].$$

Again, this is a contradiction.

By Definition I.29, Definition I.31 and Theorem I.32, then

$$L = \{\pi(s,t) \mid s = 1, 2, 3, \ldots, q, t = 1, 2, \ldots, 2^{m-1}\}$$

is an LXP-admissible partition of $\mathbb{Z}_p$ and hence it determines an LXP- neofield of order $v = p+1$.

<u>Examples</u>:

For q = 3, m = 2, we have v = 14. Take the primitive root x = 2, we then have the following 6 pairs:

$$s = 1, t = 1, \pi(1,1) = \{2^1,2^3\} = \{2,8\}$$
$$s = 1, t = 2, \pi(1,2) = \{2^2,2^4\} = \{4,3\}$$

$s = 2,\ t = 1,\ \pi(2,1) = \{2^5, 2^7\} = \{6,11\}$

$s = 2,\ t = 2,\ \pi(2,2) = \{2^6, 2^8\} = \{12,9\}$

$s = 3,\ t = 1,\ \pi(3,1) = \{2^9, 2^{11}\} = \{5,7\}$

$s = 3,\ t = 2,\ \pi(3,2) = \{2^{10}, 2^{12}\} = \{10,1\}$.

We have the presentation function of an LXP-neofield N_{14}.

x	0	1	a	a^2	a^3	a^4	a^5	a^6	a^7	a^8	a^9	a^{10}	a^{11}	a^{12}
$T(x)$	1	0	a^{10}	a^8	a^4	a^3	a^7	a^{11}	a^5	a^2	a^{12}	a	a^6	a^9

For $q = 9$, $m = 1$, then $v = 20$. Take the primitive root $x = 2$, we have the following 9 unordered pairs:

$$\{2,2^2\},\ \{2^3,2^4\},\ \ldots,\{2^{15},2^{16}\},\ \{2^{17},1\}$$

i.e.,

$$\{2,4\},\ \{8,16\},\ \{13,7\},\ \{14,9\},\ \{18,17\},\ \{15,11\},\ \{3,6\},\ \{12,5\},\ \{10,1\}.$$

This determines an LXP-neofield of order 20.

For $q = 11$, $m = 1$, we have $v = 24$. Take the primitive root $x = 5$, we have the following 11 unordered pairs:

$$\{5,5^2\},\ \{5^3,5^4\},\ \ldots,\ \{5^{19},5^{20}\},\ \{5^{21},1\}$$

i.e.,

$$\{5,2\},\ \{10,4\},\ \{20,8\},\ \{17,16\},\ \{11,9\},\ \{22,18\},\ \{21,13\},\ \{19,3\},\ \{15,6\},\ \{7,12\},\ \{14,1\}$$

This determines an LXP-neofield of order 24.

Note that the LXP-neofield N_{14} constructed above has CIP and N_{20}, N_{24} constructed are proper LXP-neofields. In this construction, the number of LXP-neofield constructed is at least $\phi(v-1)$, where ϕ is Euler's Phi-function.

Sec. 3. A Product Theorem and Examples

The purpose of this section is to construct a large family of LXP-neofields from subgroups of the group of exponents in the multiplicative group of a given order.

Let $S(\mathbf{Z}_n)$ be the symmetric group of $\mathbf{Z}_n$.

Definition IV.2: Let $\pi \in S(\mathbf{Z}_n)$. π is called an admissible permutation (or A-permutation) if

$$\{\pi(x)-x \mid x \in \mathbf{Z}_n\} = \mathbf{Z}_n .$$

Lemma IV.3: There doesn't exist any A-permutation $\pi \in S(\mathbf{Z}_n)$ such that $\pi(x) \neq x$, for every $x \in \mathbf{Z}_n$.

Proof: It follows immediately from Definition IV.2.

q.e.d.

Lemma IV.4: For any A-permutation $\pi \in S(\mathbf{Z}_n)$, there exists at most one $x \in \mathbf{Z}_n$ such that $\pi(x) = x$.

Proof: It also follows from Definition IV.2. q.e.d.

Lemma IV.5: For the positive odd integer n, we have

$$\#\{\text{A-permutation } \pi \in S(\mathbf{Z}_n) \text{ with } \pi(0) = 0\}$$
$$= \#\{\text{cyclic neofields of order } n+1\}.$$

Proof: This lemma follows from Definition I.21. q.e.d.

Theorem IV.6: For an odd number n, the total number of A-permutations of $\mathbb{Z}_n$ is $n \cdot \#\{N_{n+1}\}$, where $\#\{N_{n+1}\}$ is the number of cyclic neofields of order n+1.

Proof: Let α_k be the class of all A-permutations in $\mathbb{Z}_n$ with $\pi(k) = k$ and $\pi(x) \neq x$, for every $x \neq k$. Then by Lemma IV.5, we have

$$\#(\alpha_0) = \#\{N_{n+1}\}.$$

We will show that $\#(\alpha_k) = \#(\alpha_0)$, for all k = 1, 2, ..., n-1.

Let π be in α_0. Assume that π is as follows:

$$\pi : \begin{array}{ccccccc} 0 & 1 & 2 & \cdot & \cdot & \cdot & n-1 \\ \pi(0) & \pi(1) & \pi(2) & \cdot & \cdot & \cdot & \pi(n-1) \\ \| & & & & & & \\ 0 & & & & & & \end{array}$$

Since π is admissible, by Definition IV.2, there exists x_0 such that $\pi(x_0) = k$. Then we define $\pi_k \in \alpha_k$ to be as follows:

$$\pi_k : \begin{array}{ccccccc} 0 & \cdot\;\cdot\;\cdot & (k-1) & k & (k+1) & \ldots & (n-1) \\ \pi(x_0+n-k) & \cdot\;\cdot\;\cdot & \pi(x_0-1) & \pi(x_0) & \pi(x_0+1) & \ldots & \pi(x_0+n-k-1) \\ & & & \| & & & \\ & & & k & & & \end{array}$$

i.e., we have

$$\begin{cases} \pi_k(k) = \pi(x_o) = k \\ \pi_k(k+i) = \pi(x_o+i), \; i = 1, 2, \ldots, n-1 \pmod{n}. \end{cases}$$

Now we show that π_k is an A-permutation in α_k. It is clear that $\pi_k(y)$ are distinct, for all $y \in \mathbf{Z}_n$. Moreover, $\pi_k(k)-k = 0$ and

$$\pi_k(k+i)-(k+i) = \pi(x_o+i)-(k+i)$$

Since π is admissible in α_o, we have

$$\{\pi(x)-x \mid x \in \mathbf{Z}_n\} = \mathbf{Z}_n.$$

Hence

$$\{\pi(x_o+i)-(x_o+i) \mid i \in \mathbf{Z}_n^*\} \cup \{\pi(x_o)-x_o\} = \mathbf{Z}_n.$$

Then

$$\{\pi(x_o+i)-(k+i) \mid i \in \mathbf{Z}_n^*\} \cup \{\pi(x_o)-k\} = \mathbf{Z}_n,$$

by adding (x_o-k) to each number on left hand side of the preceding equality. Therefore, we have

$$\{\pi_k(k+i)-(k+i)\} \cup \{\pi_k(k)-k\} = \mathbf{Z}_n,$$

it follows that π_k is an A-permutation in α_k.

We will then show that two distinct A-permutations in α_o give two distinct A-permutations in α_k. Let $\pi, \pi' \in \alpha_o$ and π_k, $\pi_k' \in \alpha_k$ be as follows, with $\pi_k = \pi_k'$.

$$\pi: \begin{array}{cccccccc} 0 & 1 & 2 & 3 & \ldots & x_o & \ldots & (n-1) \\ \pi(0) & \pi(1) & \pi(2) & \pi(3) & \ldots & \pi(x_o) & \ldots & \pi(n-1) \\ \| & & & & & & & \\ 0 & & & & & & & \end{array}$$

$$\pi': \begin{array}{cccccccc} 0 & 1 & \ldots & x_o & \ldots & x_o' & \ldots & (n-1) \\ \pi'(0) & \pi'(1) & \ldots & (x_o) & & \pi'(x_o') & \ldots & \pi'(n-1) \\ 0 & & & & & & & \end{array}$$

$$\pi_k: \begin{array}{cccccc} 0 & \ldots & k & (k+1) & \ldots & (n-1) \\ \pi(x_o+n-k) & \ldots & \pi(x_o) & \pi(x_o+1) & \ldots & \pi(x_o+n-k-1) \\ & & \| & & & \\ & & k & & & \end{array}$$

$$\pi_k': \begin{array}{ccccc} 0 & \ldots & k & (k+1) & \ldots \\ \pi'(x_o'+(n-k)) & \ldots & \pi'(x_o') & \pi'(x_o'+1) & \ldots \\ \| & & \| & \| & \\ \pi(x_o+n-k) & \ldots & \pi(x_o) & \pi(x_o+1) & \ldots \\ & & \| & & \\ & & k & & \end{array}$$

Since $\pi_k = \pi_k'$, $\pi'(x_o'+i) = \pi(x_o+i)$ for all $i = n-k, \ldots, -1, 0, 1, 2, \ldots, n-1-k$, i.e., for all $i = 0, 1, 2, 3, \ldots, n-1-k, n-k, n-k+1, \ldots, n-1$. Since $\pi \in \alpha_o$, $\pi(0) = 0$. We then have two cases, $x_o' \neq x_o$ and $x_o' = x_o$.

If $x_o' \neq x_o$, then without loss of generality we assume $x_o' > x_o$. Hence

$$\begin{aligned} \pi'(x_o'-x_o) &= \pi'(x_o'+(-x_o)) \\ &= \pi(x_o+(-x_o)) \\ &= \pi(0) \\ &= 0. \end{aligned}$$

It follows that there exists $x_o'-x_o \neq 0$ in $\mathbb{Z}_n$ such that $\pi'(x_o'-x_o) = 0$, this contradicts to $\pi' \in \alpha_o$. If $x_o' = x_o$, then $\pi' = \pi$. Therefore we have

$$\#(\alpha_o) \leq \#(\alpha_k).$$

Conversely, we have $\#(\alpha_k) \leq \#(\alpha_o)$. Hence $\#(\alpha_k) = \#(\alpha_o)$, for all $k = 1, 2, \ldots, n-1$. By Lemma IV.3 and Lemma IV.4, all the A-permutations are in $\bigcup_{i=0}^{n-1} \alpha_i$. Hence by Lemma IV.5, the total number of A-permutations in $S(\mathbb{Z}_n)$ is $n \cdot \#(\alpha_o) = n \cdot \#(N_{n+1})$. q.e.d.

Theorem IV.7: For v even and $G = \mathbb{Z}_{v-1}$; let H be a subgroup of G, L_H and $L_{G/H}$ be LXP-neofields defined on H and G/H respectively. Let π be an A-permutation of H. Let $(H+x, H+y)$ be coset representatives for each pair of $L_{G/H}$, pair these cosets with

$$\left\{(h+x, \pi(h)+y) \mid h \in H\right\}.$$

Let L_G be the union of L_H and a set of such pairs for each coset pair of $L_{G/H}$. Then L_G is an LXP-neofield of order v.

Proof: Let

$$H = \left\{0, \frac{v-1}{n}, 2\,\frac{v-1}{n}, 3\,\frac{v-1}{n}, \ldots, (n-1)\,\frac{v-1}{n}\right\}.$$

Then $G/H = \left\{H, H+1, H+2, \ldots H+\frac{v-1}{n} - 1\right\}$. Let

$$\begin{cases} L_H = \left\{(x_i, y_i) \mid x_i, y_i \in H,\ 1 \leq i \leq r\right\}, \text{ where } r = \frac{n-1}{2}. \\ L_{G/H} = \left\{(H+a_j, H+b_j) \mid 1 \leq j \leq s\right\}, \text{ where } s = \frac{v-1-n}{2n}. \end{cases}$$

Then

$$L_G = L_H \cup \{(h+a_j, \pi(h)+b_j) \mid h \text{ runs over } H, 1 \leq j \leq s\}.$$

Since L_H is an LXP-neofield, $\{x_i\} \cup \{y_i\} = H^*$ and

$$\{\pm(x_i - y_i) \mid 1 \leq i \leq r\} = H^* \quad , \text{ where } H^* = H - \{0\}.$$

We then have

$$\#\{x_i, y_i \mid (x_i, y_i) \in L_H\} = 2r = 2\,\frac{n-1}{2} = (n-1).$$

In L_G, $\{\pi(h)+b_j\} \cap \{h+a_j\} = \emptyset$, since they belong to different cosets in G/H. Moreover,

$$\{h+a_j, \pi(h)+b_j\} \cap \{x_i, y_i \mid (x_i, y_i) \in L_H\} = \emptyset,$$

since $(H+i) \cap H = \emptyset$. Therefore

$$\#\{x, y \mid (x, y) \in L_G\} = 2r + 2s \cdot \#(H) = v-2$$

Hence $\{x, y \mid (x, y) \in L_G\} = G^*$.

Next, we show that all the differences in L_G are distinct. Since

$$\{\pm(x_i - y_i) \mid (x_i, y_i) \in L_H\} = H^*,$$

we have

$$\#\{\pm(x_i - y_i) \mid (x_i, y_i) \in L_H\} = n-1.$$

Let $d_j = (\pi(h)+b_j) - (h+a_j)$

$= (\pi(h)-h) + (b_j - a_j)$.

Then for two distinct h_1, h_2 (still fix j), we have

$$\begin{cases} d_j^{(1)} = (\pi(h_1)-h_1)+(b_j-a_j) \\ d_j^{(2)} = (\pi(h_2)-h_2)+(b_j-a_j). \end{cases}$$

Since π is an A-permutation, $\pi(h_1)-h_1 \neq \pi(h_2)-h_2$ if $h_1 \neq h_2$. Hence $d_j^{(1)} \neq d_j^{(2)}$. Now let j vary, we have

$$\begin{cases} d_j^{(1)} = (\pi(h_1)-h_1)+(b_{j_1}-a_{j_1}) \\ d_j^{(2)} = (\pi(h_2)-h_2)+(b_{j_2}-a_{j_2}) \end{cases}$$

Then

$$\begin{cases} d_j^{(1)} \equiv (b_{j_1}-a_{j_1}) \pmod{\frac{v-1}{n}} \\ d_j^{(2)} \equiv (b_{j_2}-a_{j_2}) \pmod{\frac{v-1}{n}} \end{cases}$$

But since $(H+a_{j_1}, H+b_{j_1})$ and $(H+a_{j_2}, H+b_{j_2})$ are two pairs in $L_{G/H}$, $b_{j_1}-a_{j_1} \not\equiv b_{j_2}-a_{j_2} \pmod{\frac{v-1}{n}}$. Hence

$$d_j^{(1)} \not\equiv d_j^{(2)} \pmod{\frac{v-1}{n}}.$$

Therefore

$$d_j^{(1)} \not\equiv d_j^{(2)} \pmod{(v-1)}.$$

Since $d_j^{(i)} = (\pi(h_i)-h_i)+(b_j-a_j) \not\equiv 0 \pmod{\frac{v-1}{n}}$, we have

$$\begin{aligned} \#\left\{\pm(x-y) \,\middle|\, (x,y)\in L_G\right\} &= (n-1)+2\cdot s\cdot \#(H) \\ &= v-2, \end{aligned}$$

i.e., $\left\{\pm(x,y) \,\middle|\, (x,y)\in L_G\right\} = G^*$. Combining this with the previous paragraph, L_G is an LXP-neofield. <u>q.e.d.</u>

Theorem IV.8: As in Theorem IV.7, there are at least $f \cdot t \cdot a^s$ LXP-neofields of order v associated with the subgroup H of G, where

$$\begin{cases} f = \#(\text{LXP-neofields in } H) \\ t = \#(\text{LXP-neofields in } G/H) \\ a = \#(\text{A-permutations of } H) \\ s = \dfrac{\#(G/H)-1}{2} \end{cases}$$

Proof: We write

$$L_G = L_H \cup \{(h+x, \pi(h)+y) \mid h \text{ runs over } H\}.$$

For any other notations, we follow Theorem IV.7.

To choose L_H, we have f possibilities, where f = #(LXP-neofields in H). Now if we want to construct L_G from a given LXP-neofield in G/H and a given LXP-neofield in H. Let π be any A-permutation of H, we have

$$\{(h+x, \pi(h+y)\} = T_1 \cup T_2 \cup \ldots \cup T_s,$$

where $s = \frac{\#(G/H)-1}{2}$ is the number of pairs in the given LXP-neofield in G/H and T_m is the set of pairs associated with m-th pair in $L_{G/H}$. If π changes, the above decomposition changes. Hence we have

$$\{(h+x, \pi_i(h)+y)\} = T_{i_1} \cup T_{i_2} \cup \ldots \cup T_{i_s}.$$

Note that $T_{i_h} \cap T_{i_k} = \emptyset$ if $h \neq k$, since the numbers in T_{i_h} and T_{i_k} do belong to different cosets of G/H. Also we have that T_{i_m} has n elements for each i and m = 1, 2, ..., s, since each coset in G/H has n elements. For some h, $1 \leq h \leq s$, we have

$$T_{i_h} \cap T_{j_h} = \emptyset,$$

where T_{i_h} and T_{j_k} are associated with the A-permutation π_i and π_j respectively. The first elements in pairs of T_{i_h} are the same as those of T_{j_k}, but π_i, π_j are distinct A-permutations, which send the same first element to distinct second elements. Therefore we have the following possibilities for $\{(h+x, \pi(h)+y)\}$:

$$T_{11} \cup T_{12} \cup \ldots \cup T_{1s}$$
$$T_{21} \cup T_{22} \cup \ldots \cup T_{2s}$$
$$\cdot$$
$$\cdot$$
$$\cdot$$
$$T_{a1} \cup T_{a2} \cup \ldots \cup T_{as}$$

Diagram IV.1

where a is the number of A-permutations of H.

After applying A-permutations, the set of differences of the pairs in T_{i_h} and in T_{j_h} are disjoint but equal (mod $\frac{v-1}{n}$). Hence to construct an LXP-neofield, we are free to choose one set in each column of Diagram IV.1 and combine them together to get $\{(h+x, \pi_i(h)+y)\}$ and then together with L_H, we get an LXP-neofield. Therefore, we have a^s ways. Then in total, we have at least

$$f \cdot t \cdot a^s$$

choices, where f = # of LXP-neofields in H and t = # of LXP-neofields in G/H. This completes the proof of the theorem. <u>q.e.d.</u>

<u>Example</u>: Take v = 22. Then $G = \mathbb{Z}_{21}$. We divide into two cases.

<u>Case 1</u>: Let $H = \{0, 7, 14\} \cong \mathbb{Z}_3$. Then

$$G/H = \{H, H+1, \ldots, H+6\} \cong \mathbb{Z}_7.$$

LXP-neofield in H: (7,14)

LXP-neofield in G/H: (a). (13) (26) (45)

(b). (15) (46) (23)

(c). (16) (25) (34)

(A-permutations of H) = $3 \cdot \#(N_4) = 3 \cdot 1 = 3$, i.e.,

$$\pi_0 : \begin{pmatrix} 0 & 7 & 14 \\ 0 & 14 & 7 \end{pmatrix}, \quad \pi_1 : \begin{pmatrix} 0 & 7 & 14 \\ 14 & 7 & 0 \end{pmatrix}, \quad \pi_2 : \begin{pmatrix} 0 & 7 & 14 \\ 7 & 0 & 14 \end{pmatrix}$$

Hence we have 27 LXP-neofields associated with(a) , i.e.,

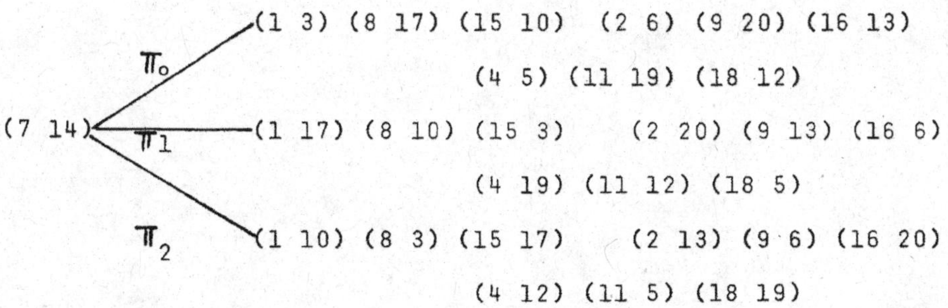

and 27 LXP-neofields associated with each of (b) and (c). In this case, we obtain $f \cdot t \cdot a^s = 1 \cdot 3 \cdot 3^3 = 81$ LXP-neofields.

Case 2: Let $H = \{0, 3, 6, 9, 12, 15, 18\} \cong \mathbb{Z}_7$. Then

$$G/H = \{H, H+1, H+2\} \cong \mathbb{Z}_3.$$

$H+1 = \{1, 4, 7, 10, 13, 16, 19\}$

$H+2 = \{2, 5, 8, 11, 14, 17, 20\}$.

Here we have
$$\begin{cases} f = \#(\text{LXP-neofields in } H) \\ \quad = \#(\text{LXP-neofields of order } 8) \\ \quad = 3. \\ t = \#(\text{LXP-neofields in } G/H) \\ \quad = 1 \\ a = \#(\text{A-permutations of } H),\ H \cong \mathbb{Z}_7 \\ \quad = 7 \cdot \#(N_8) \\ \quad = 7 \cdot 19 \\ \quad = 133 \\ s = \frac{3-1}{2} = 1. \end{cases}$$

Hence the total number of LXP-neofields in this case = $3 \cdot 1 \cdot (133)^1 = 399$. We have the following:

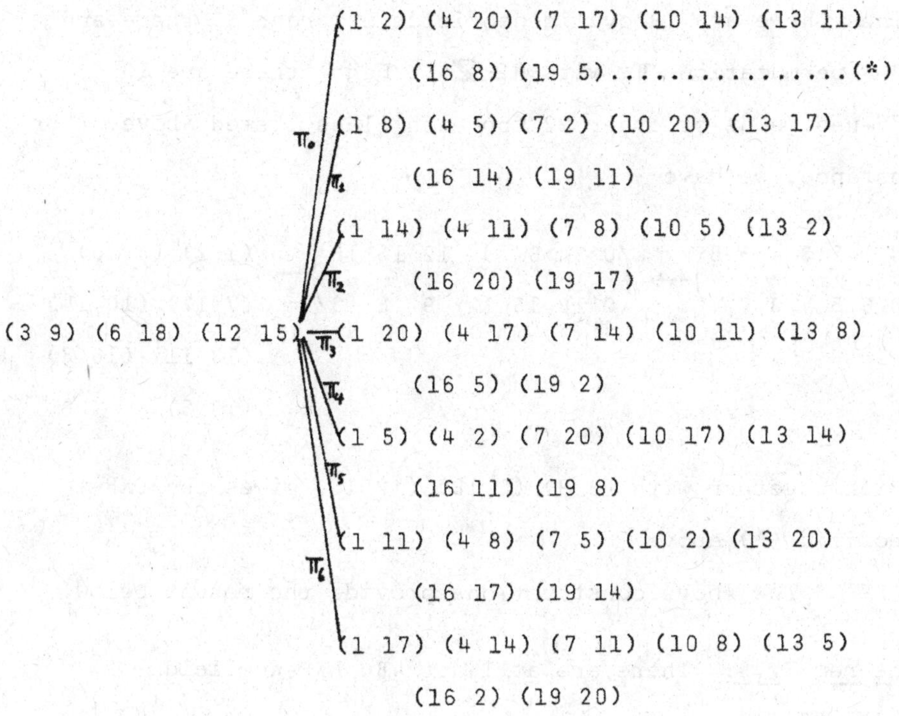
(1 2) (4 20) (7 17) (10 14) (13 11)
(16 8) (19 5)...............(*)
π_0
(1 8) (4 5) (7 2) (10 20) (13 17)
π_1
(16 14) (19 11)
(1 14) (4 11) (7 8) (10 5) (13 2)
π_2
(16 20) (19 17)
(3 9) (6 18) (12 15)
(1 20) (4 17) (7 14) (10 11) (13 8)
π_3
(16 5) (19 2)
π_4
(1 5) (4 2) (7 20) (10 17) (13 14)
π_5
(16 11) (19 8)
(1 11) (4 8) (7 5) (10 2) (13 20)
π_6
(16 17) (19 14)
(1 17) (4 14) (7 11) (10 8) (13 5)
(16 2) (19 20)

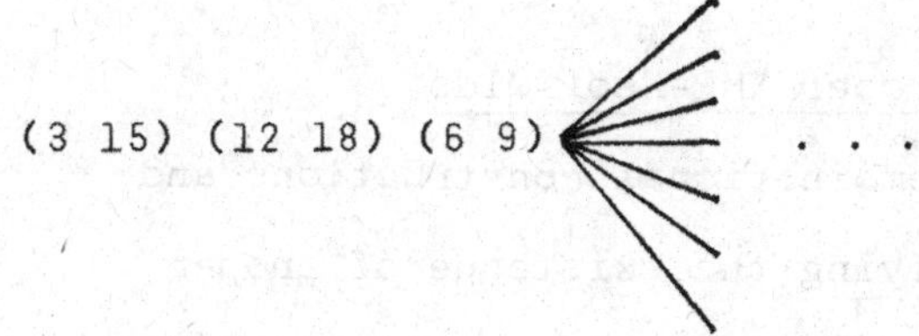
(3 15) (12 18) (6 9)
. . .

(3 18) (6 15) (9 12) . . .

Since there are 19 cyclic neofields of order 8, there are 19 A-permutations π_0 with $H \cong \mathbb{Z}_7$. Hence there are 19 LXP-neofields of order 22 for each class listed above. For instance, we have

$$\begin{pmatrix} 0 & 1 & 2 & 3 & 4 & 5 & 6 \\ 0 & 6 & 5 & 4 & 3 & 2 & 1 \end{pmatrix} \longrightarrow \begin{pmatrix} 0 & 3 & 6 & 9 & 12 & 15 & 18 \\ 0 & 18 & 15 & 12 & 9 & 6 & 3 \end{pmatrix} \longrightarrow \begin{matrix} (1\ 2)\ (4\ 20) \\ (7\ 17)\ (10\ 14) \\ (13\ 11)\ (16\ 8) \\ (19\ 5) \end{matrix}$$

which together with (3 9) (6 18) (12 15) gives the LXP-neofield (*) above

The above constructions provide the result below.

Theorem IV.9: There are at least 480 LXP-neofields of order 22.

Sec. 4. Construction of Proper XMP-Neofields

In this section, combinational constructions and examples are provided verifying the existence of proper XMP-neofields (a proper XMP-neofield is a cyclic neofield without any of the properties XIP, LXP, CMP or RXP) of all orders $v \geq 7$.

Case 1: $v = 4m$, $m \geq 2$.

We partition $\mathbb{Z}^*_{4m-1}$ into 4m-2 ordered pairs $\{k,n\}$ which gives an XMP-admissible partition of $\mathbb{Z}^*_{4m-1}$.

We pick k and n by the following scheme XA1:

k	n	n-k
1	2m-1	2m-2
2	2m-2	2m-4
.	.	.
.	.	.
.	.	.
m-1	m+1	2
m+1	m	4m-2
m+2	m-1	4m-4
.	.	.
.	.	.
.	.	.
2m-1	2	2m+2

XA1

yielding the differences $\{n-k\}$ all the even numbers from 2 to 4m-2 (except 2m).

Next, we pair k and n by scheme XA2.

k	n	n-k
2m+1	4m-2	2m-3
2m+2	4m-3	2m-5
.	.	.
.	.	.
.	.	.
3m-1	3m	1
3m	3m-2	4m-3
3m+1	3m-3	4m-5

(continued)

k	n	n-k
.	.	.
.	.	.
.	.	.
4m-2	2m	2m+1

XΛ2

yielding the differences $\{n-k\}$ all the odd numbers from 1 to 4m-3 (except 2m-1).

Then we have the following two ordered pairs

k	n	n-k
m	3m-1	2m-1
2m	1	2m

giving the differences $\{n-k\} = \{2m, 2m-1\}$.

One checks that this is an XMP-admissible partition of $\mathbf{Z}^*_{4m-1}$. By Theorem I.23, this determines a proper XMP-neofield of order v = 4m.

Examples:

For m = 3, v = 12. We have the ordered pairs:

k	n	n-k
1	5	4
2	4	2
4	3	10
5	2	8
- - -	- - -	- - -
7	10	3

(continued)

k	n	n-k
8	9	1
10	6	7
9	7	9
- - -	- - -	- - -
3	6	5
6	1	6

For m = 2, v = 8. We have the ordered pairs:

k	n	n-k
1	3	2
3	2	6
- - -	- - -	- - -
5	6	1
6	4	5
- - -	- - -	- - -
2	5	3
4	1	4

Case 2: v = 4m+2, m $\geq$ 2.

We partition $\mathbf{Z}^*_{4m+1}$, into 4m ordered pairs $\{k,n\}$ which gives an XMP-admissible partition of $\mathbf{Z}^*_{4m+1}$.

We pick k and n by scheme XA3

k	n	n-k
1	2m	2m-1
2	2m-1	2m-3
.	.	.
.	.	.
.	.	.
m	m+1	1
m+2	m	4m-1
m+3	m-1	4m-3
.	.	.
.	.	.
.	.	.
2m	2	2m+3

XA3

yielding the differences $\{n-k\}$ all the odd elements from 1 to 4m-1 (except 2m+1).

Next, we pair k and n by scheme XA4:

k	n	n-k
2m+2	4m	2m-2
2m+3	4m-1	2m-4
.	.	.
.	.	.
.	.	.
3m	3m+2	2
3m+1	3m	4m
3m+2	3m-1	4m-2

(continued)

k	n	$n-k$
.	.	.
.	.	.
.	.	.
$4m$	$2m+1$	$2m+2$

yielding the differences $\{n-k\}$ all the even elements from 2 to $4m$ (except $2m$) .

Then we have the following two ordered pairs:

k	n	$n-k$
$m+1$	$3m+1$	$2m$
$2m+1$	1	$2m+1$

giving the differences $\{n-k\} = \{2m, 2m+1\}$.

One checks that this is an XMP-admissible partition of $\mathbb{Z}^*_{4m+1}$. By Theorem I.23 again, this determines a proper XMP-neofield of order $4m+2$.

Examples:

For $m = 2$, $v = 10$. We have $\{k, n, n-k\}$:

k	n	$n-k$
1	4	3
2	3	1
4	2	7
- - -	- - -	- - -
6	8	2
7	6	8

(continued)

k	n	n-k
8	5	6
- - -	- - -	- - -
3	7	4
5	1	5

of proper XMP-neofield of order 10.

For m = 3, v = 14. We have $\{k, n, n-k\}$:

k	n	n-k
1	6	5
2	5	3
3	4	1
5	3	11
6	2	9
- - -	- - -	- - -
8	12	4
9	11	2
10	9	12
11	8	10
12	7	8
- - -	- - -	- - -
4	10	6
7	1	7

of a proper XMP-neofield of order 14.

Case 3: $v = 4m+1$, $m \geq 2$.

We partition $\mathbf{Z}_{4m}^*$ into 4m-2 ordered pairs $\{k,n\}$ and a singleton $\{2m+1\}$ which give an XMP-admissible partition of $\mathbb{Z}_{4m}^*$.

We pick k and n by scheme XB1:

k	n	n-k
0	2m-1	2m-1
1	2m-2	2m-3
.	.	.
.	.	.
.	.	.
m-1	m	1
4m-1	2m+2	2m+3
4m-2	2m+3	2m+5
.	.	.
.	.	.
.	.	.
3m+1	3m	4m-1

XB1

yielding the differences $\{n-k\}$ all the odd elements from 1 to 4m-1 (except 2m+1).

Next, we pair k and n by scheme XB2:

k	n	n-k
2m-1	1	2m+2
2m-2	2	2m+4
.	.	.
.	.	.
.	.	.

(continued)

k	n	n-k
m+1	m-1	4m-2
2m+2	0	2m-2
2m+3	4m-1	2m-4
.	.	.
.	.	.
.	.	.
3m	3m+2	2

XB2

yielding the differences $\{n-k\}$ all the even elements from 2 to 4m-2 (except 2m).

Then we have the following two ordered pairs:

k	n	n-k
m	3m+1	2m+1
2m+1	2m+1	0

giving the differences 0 and 2m+1.

One checks that this is an XMP-admissible partition $\mathbb{Z}_{4m}^*$. By Theorem I.23, this determines a proper XMP-neofield of order 4m+1.

Examples:

For m = 2, v = 9. We have the ordered pairs and a singleton $\{5\}$:

k	n	n-k
0	3	3
1	2	1
7	6	7
- - -	- - -	- - -
3	1	6
6	0	2
- - -	- - -	- - -
2	7	5
5	5	0

for a proper XMP-neofield of order 9.

For m = 3, v = 13. We have the ordered pairs and a singleton $\{7\}$:

k	n	n-k
0	5	5
1	4	3
2	3	1
10	9	11
11	8	9
- - -	- - -	- - -
5	1	8
4	2	10
8	0	4
9	11	2
- - -	- - -	- - -

(continued)

k	n	n-k
3	10	7
7	7	0

of a proper XMP-neofield of order 13.

Case 4: $v = 4m+3$, $m \geq 1$.

We partition $\mathbf{Z}^*_{4m+2}$ into 4m ordered pairs $\{k,n\}$ and a singleton $\{2m+2\}$ which give an XMP-admissible partition of $\mathbf{Z}^*_{4m+2}$.

We pick k and n by scheme XB3:

k	n	n-k
0	2m	2m
1	2m-1	2m-2
.	.	.
.	.	.
.	.	.
m-1	m+1	2
4m+1	2m+3	2m+4
4m	2m+4	2m+6
.	.	.
.	.	.
.	.	.
3m+3	3m+1	4m

XB3

yielding differences $\{n-k\}$ all even elements from 2 to 4m (except 2m+2).

Next, we pair k and n by scheme XB4:

k	n	n-k
2m	1	2m+3
2m-1	2	2m+5
.	.	.
.	.	.
.	.	.
m+1	m	4m+1
3m+2	3m+3	1
3m+1	3m+4	3
.	.	.
.	.	.
.	.	.
2m+3	0	2m-1

XB4

yielding differences $\{n-k\}$ all odd elements from 1 to 4m+1 (except 2m+1).

Then we have the following two pairs:

k	n	n-k
m	3m+2	2m+2
2m+2	2m+2	0

giving the differences 0 and 2m+2.

One checks that this is an XMP-admissible partition of $\mathbf{Z}^*_{4m+2}$. By Theorem I.23, this gives a proper XMP-neofield of order 4m+3.

Examples:

For $m = 1$, $v = 7$. We have $\{k, n, n-k\}$:

k	n	n-k
0	2	2
5	0	1
2	1	5
- - -	- - -	- - -
1	5	4
4	4	0

of a proper XMP-neofield N_7.

For $m = 2$, $v = 11$. We have $\{k, n, n-k\}$:

k	n	n-k
0	4	4
1	3	2
9	7	8
- - -	- - -	- - -
4	1	7
3	2	9
8	9	1
7	0	3
- - -	- - -	- - -
2	8	6
6	6	0

of a proper XMP-neofield of order 11.

For $m = 3$, $v = 15$. We have $\{k, n, n-k\}$:

k	n	n-k
0	6	6
1	5	4
2	4	2
12	10	12
13	9	10
- - -	- - -	- - -
6	1	9
5	2	11
4	3	13
11	12	1
10	13	3
9	0	5
- - -	- - -	- - -
3	11	8
8	8	0

of a proper XMP-neofield of order 15.

The above four cases give the result below.

<u>Theorem IV.10</u>: There exist proper XMP-neofield for any order $v \geq 7$.

CHAPTER V

CYCLIC NEOFIELDS AND COMBINATORIAL DESIGNS

Sec. 1. Triple Systems Arising from XIP-Neofields

Johnsen [14] has pointed out that existence of a CIP-neofield of even order v is equivalent to the existence of a cyclic Steiner triple system of order v-1. Doner [4] has shown that the existence of an odd order CIP-neofield is equivalent to the existence of an "almost" cyclic Steiner triple system. In this section, we show that XIP-neofields give rise to BIBD's with $\lambda = 2$.

Let S be a set with n elements and $\mathcal{B}$ a collection of distinct k-subsets of S. Then we have:

Definition V.1: $(S, \mathcal{B})$ is called a (n, b, r, k, λ)-balanced incomplete block design (or (n, b, r, k, λ)-BIBD) if

i) $|\mathcal{B}| = b$,

ii) every element of S appears in exactly r of the k-subsets in $\mathcal{B}$,

iii) every pair of two distinct elements of S appears as a subset of λ of the k-subsets of $\mathcal{B}$,

iv) $0 < \lambda, k < n-1$.

The k-subsets in $\mathcal{B}$ are called "blocks" of the BIBD. Moreover, it is well-known that $bk = nr$ and $r\cdot(k-1) = \lambda \cdot(n-1)$ for a (n, b, r, k, λ)-BIBD.

<u>Definition V.2</u>: An XIP-neofield is said to have <u>local CIP-property</u> if any of the defining relations XIA or XIB (Definitions I.45 and I.58) on the sextuples of an XIP-admissible partition is of type CIA or CIB respectively.

Note that every XIP-neofield of type a) i) or type a) iii) has local CIP. For an XIP-neofield of type b), it has local CIP when $t \neq 0$ and XIA = CIA (or CIB) for any of the isolated sextuples in H_t.

<u>Definition V.3</u>: A proper XIP-neofield is said to be <u>totally proper</u> if it doesn't have local CIP.

<u>Definition V.4</u>: A (n, b, r, k, λ)-BIBD, $(\mathbb{Z}_n, \mathcal{B})$ is called <u>semi-cyclic</u> if $\mathcal{B}$ can be divided into equivalence classes $\mathcal{B}_i$ such that each class has the same number of blocks and each class is cyclic in the sense that $\{r,s,t\} \in \mathcal{B}_i \Leftrightarrow \{r+j, s+j, t+j\} \in \mathcal{B}_i$ for any $j \in \mathbb{Z}_n$.

From Section I.1, the addition table of every cyclic neofield is characterized by its presentation function. So for an XIP-neofield N_v, let T be defined as follows:

$$\begin{cases} \text{i)} & \{0,i,j\} \in T \text{ if and only if } 1+a^i = \begin{cases} a^j & \text{if } v \text{ is even.} \\ a^{j+\ell} & \text{if } v \text{ is odd.} \end{cases} \\ \text{ii)} & \{i,j,k\} \in T \text{ if and only if } \{i+1,\ j+1,\ k+1\} \pmod{v-1} \in T. \end{cases}$$

<u>Theorem V.5</u>: For $v \equiv 2 \pmod 6$ and a totally proper XIP-neofield N_v, the triple system $(\mathbb{Z}_{v-1}, T)$ defined above is a $(v-1,\ \frac{(v-1)(v-2)}{3},\ v-2,\ 3,\ 2)$-BIBD.

Proof: It is obvious that $\mathbb{Z}_{v-1}$ has v-1 elements and each block in T has 3 elements.

For each block of the type $\{0,k,n\}$ from $1+a^k = a^n$ in N_v, we have v-2 more blocks $\{0+i, k+i, n+i\}$ (mod v-1) in T, where i = 1, 2, 3, ..., v-2. We claim that these v-1 blocks are distinct. It is sufficient to show that two different blocks have at most one element in common. Without loss of generality, we assume $k+i_1 = 0+i_2$ and $0+i_1 = n+i_2$. Hence $i_2-i_1 \equiv k \pmod{v-1}$ and $i_2-i_1 \equiv -n \pmod{v-1}$, i.e., $k \equiv -n \pmod{v-1}$. This is impossible since N_v has XIP and (k,n) is admissible in the sense of Definition I.36. Let T(k,n) be the equivalent class in T consisting of blocks generated by the initial block $\{0,k,n\}$.

We have v-2 elements in the second row (excluding 0 and 1) of the addition table. Hence

$$T = \bigcup_{(k,n)} T(k,n)$$

with $|T(k,n)| = v-1$ for each (k,n). Therefore, we have (v-1)(v-2) triples in T counting multiplicities.

By Lemma I.15 a), T(n-k,-k) and T(-n,k-n) are subsets of T whenever T(k,n) is. Moreover,

$$\{0,n-k,-k\} \in T(n-k,-k)$$
$$\Rightarrow \{k,n,0\} \in T(n-k,-k)$$

and

$$\{0,k,n\} \in T(k,n)$$
$$\Rightarrow \{-k,0,n-k\} \in T(k,n).$$

Then $T(k,n) = T(n-k,-k)$. Similarly, we have $T(k,n) = T(-n,k-n)$. Therefore, we have $\frac{(v-1)(v-2)}{3}$ blocks in T and

$$T = \bigcup_{(k,n)} T(k,n) = \bigcup_{r=1}^{e} T_r, \text{ where } e = \frac{v-2}{3}.$$

Each element of $\mathbb{Z}_{v-1}$ appears in three blocks in an equivalence class T_r, hence it appears in $3e = (v-2)$ blocks of T. Now let $\{s,t\}$ be a pair of distinct elements from $\mathbb{Z}_{v-1}$. We assume the following defining relation in the addition table of N_v:

$$\begin{cases} a^s+a^t = a^{\ell_1}, \\ a^{\ell_2}+a^s = a^t, \\ a^s+a^{\ell_3} = a^t, \end{cases} \qquad \begin{cases} a^t+a^s = a^{\ell_4}, \\ a^{\ell_5}+a^t = a^s, \\ a^t+a^{\ell_6} = a^s, \end{cases}$$

with $\ell_1 \neq \ell_4$, $\ell_2 \neq \ell_5$ and $\ell_3 \neq \ell_6$, since N_v is totally proper. Then we have:

$$\begin{aligned} a^s+a^t = a^{\ell_1} &\Rightarrow 1+a^{t-s} = a^{\ell_1-s} \\ &\Rightarrow 1+a^{s-\ell_1} = a^{t-\ell_1} \quad \text{(by a) iii) of Lemma I.15)} \\ &\Rightarrow a^{\ell_1}+a^s = a^t \end{aligned}$$

Hence $\ell_1 \equiv \ell_2 \pmod{v-1}$. Similarly, $\ell_1 \equiv \ell_3 \pmod{v-1}$. Thus $\ell_1 \equiv \ell_2 \equiv \ell_3 \pmod{v-1}$. By the same argument, we have $\ell_4 \equiv \ell_5 \equiv \ell_6 \pmod{v-1}$. Hence

$$\{s,t\} \subset \{s,t,\ell_1\} = \{\ell_1,s,t\} = \{s,\ell_1,t\} \in T(\ell_1-s\ ,t-s)$$

and

$$\{s,t\} \subset \{t,s,\ell_4\} = \{\ell_4,t,s\} = \{t,\ell_4,s\} \in T(\ell_4-t,s-t)$$

Therefore we have $\lambda = 2$.

Summarizing the above arguments, we conclude that $(\mathbb{Z}_{v-1},T)$ is a $(v-1, \frac{(v-1)(v-2)}{3}, v-2, 3, 2)$-BIBD. q.e.d.

Corollary V.6: The triple system $(\mathbb{Z}_{v-1},T)$ derived from a totally proper XIP-neofield of order $v \equiv 2 \pmod 6$ is semi-cyclic.

Definition V.7: A set $B = \{0,k,n\}$ from $\mathbb{Z}_m$, m odd, is said to be admissible if the pair (k,n) (or the pair (n,k)) is admissible (mod m) in the sense of Definition I.36.

Definition V.8: A semi-cyclic (n, b, r, k,λ)-BIBD, $(\mathbb{Z}_n,\beta)$, is said to be admissible if the initial blocks (the blocks with one zero) of equivalence classes in β are admissible.

Corollary V.9: The triple system $(\mathbb{Z}_{v-1},T)$ in Theorem V.5 is admissible.

The above corollary indicates that a necessary condition that a totally proper XIP-neofield of even order v exists is that there exists an admissible $(v-1, \frac{(v-1)(v-2)}{3}, v-2, 3, 2)$-BIBD. We will show that this condition is also sufficient in Section V.4. Now we give two examples to illustrate BIBD's arising from totally proper XIP-neofield of order $v \equiv 2 \pmod 6$.

Examples:

For v = 14, we have a totally proper XIP-neofield of type b):

x	0	1	a	a^2	a^3	a^4	a^5	a^6	a^7	a^8	a^9	a^{10}	a^{11}	a^{12}
$1+x$	1	0	a^4	a^6	a^{12}	a^{11}	a^2	a	a^9	a^7	a^{10}	a^8	a^3	a^5 .

We then have the XIP-admissible partition of type b) of $\mathbb{Z}_{13}^*$.

$$S(1,4) = \{1,\ 4,\ 3,\ 12,\ 9,\ 10\}$$
$$S(2,6) = \{2,\ 6,\ 4,\ 11,\ 7,\ 9\}$$
$$S(5,2) = \{5,\ 2,\ 10,\ 8,\ 11,\ 3\}$$
$$S(6,1) = \{6,\ 1,\ 8,\ 7,\ 12,\ 5\}$$

Hence we have:

$$T(1,4) = \{\{0,1,4\},\ \{1,2,5\},\ \ldots,\ \{12,0,3\}\}$$
$$T(2,6) = \{\{0,2,6\},\ \{1,3,7\},\ \ldots,\ \{12,1,5\}\}$$
$$T(5,2) = \{\{0,5,2\},\ \{1,6,3\},\ \ldots,\ \{12,4,1\}\}$$
$$T(6,1) = \{\{0,6,1\},\ \{1,7,2\},\ \ldots,\ \{12,5,0\}\}.$$

Let $T = T(1,4) \cup T(2,6) \cup T(5,2) \cup T(6,1)$. One checks that $(\mathbb{Z}_{13}, T)$ is an admissible semi-cyclic (13, 52, 12, 3, 2)-BIBD.

For $v = 14$, we have a totally proper XIP-neofield of type a) ii):

x	0	1	a	a^2	a^3	a^4	a^5	a^6	a^7	a^8	a^9	a^{10}	a^{11}	a^{12}
$1+x$	1	0	a^{10}	a^8	a^4	a	a^7	a^{11}	a^2	a^6	a^{12}	a^9	a^5	a^3

We then have the following type a) XIP-admissible partition of $\mathbb{Z}_{13}^*$.

$$S(3,4) = \{3,\ 4,\ 1,\ 10,\ 9,\ 12\}$$
$$S(5,7) = \{5,\ 7,\ 2,\ 8,\ 6,\ 11\}.$$

Hence we have:

$$T(3,4) = \{\{0,3,4\}, \{1,4,5\}, \ldots, \{12,2,3\}\}$$
$$T(4,1) = \{\{0,4,1\}, \{1,5,2\}, \ldots, \{12,3,0\}\}$$

$$T(5,7) = \{\{0,5,7\}, \{1,6,8\}, \ldots, \{12,4,6\}\}$$
$$T(7,2) = \{\{0,7,2\}, \{1,8,3\}, \ldots, \{12,6,1\}\}.$$

and let

$$T = T(3,4) \cup T(4,1) \cup T(5,7) \cup T(7,2).$$

One can verify that $(\mathbb{Z}_{13}, T)$ is an admissible semi-cyclic (13, 52, 12, 3, 2)-BIBD.

We now turn our attention to the case when $v \equiv 4 \pmod 6$. For an XIP-neofield of order $v \equiv 4 \pmod 6$, we always have a pair $\{k,n\} = \{\frac{v-1}{3}, \frac{2(v-1)}{3}\}$ in an XIP-admissible partition of $\mathbb{Z}^*_{v-1}$. The remaining sextuples are similar to the case when $v \equiv 2 \pmod 6$.

<u>Lemma V.10</u>: For $v \equiv 4 \pmod 6$ and the special pair $\{k,n\} = \{\frac{v-1}{3}, \frac{2(v-1)}{3}\}$, $T(k,n)$ consists of $\frac{v-1}{3}$ distinct triples, and no pair in $\mathbb{Z}_{v-1} \times \mathbb{Z}_{v-1}$ is contained in more than one triple.

<u>Proof</u>: Let $a = \frac{v-1}{3}$. Since $\{0,k,n\} = \{0,a,2a\}$, we have

$$\{0,a,2a\} = \{0+a,\ a+a,\ 2a+a\} = \{0+2a,\ a+2a,\ 2a+2a\}.$$

Hence there are $a = \frac{v-1}{3}$ distinct triples in $T(k,n)$, i.e.,

$$\{0,a,2a\},\ \{1,a+1,2a+1\},\ \ldots,\ \{a-1,a+(a-1),2a+(a-1)\}.$$

The remaining part of the lemma is obvious. <u>q.e.d.</u>

<u>Definition V.11</u>: Let $T_o = T(\frac{v-1}{3}, \frac{2(v-1)}{3})$ be the collection of $\frac{v-1}{3}$ triples generated by $\left\{0, \frac{v-1}{3}, \frac{2(v-1)}{3}\right\}$.

<u>Definition V.12</u>: A BIBD, (S,β), is said to be <u>hyper-semi-cyclic</u> if β can be divided into equivalence class β_i such that each class has the same number (say v-1) of blocks, except one class having r blocks with $r < v-1$, and each class is cyclic in the sense of Definition V.3. A hyper-semi-cyclic BIBD with all initial blocks admissible is said to be an <u>admissible hyper-semi-cyclic BIBD</u>.

<u>Theorem V.13</u>: For $v \equiv 4 \pmod 6$ and a totally proper XIP-neofield N_v, the triple system $(\mathbb{Z}_{v-1},T)$ derived from N_v with T_o repeated once is an admissible hyper-semi-cyclic BIBD with parameters $(v-1, \frac{(v-1)(v-2)}{3}, v-2, 3, 2)$.

<u>Proof</u>: That $(\mathbb{Z}_{v-1},T)$ is admissible is clear, since N_v has an XIP-admissible partition. That $(\mathbb{Z}_{v-1},T)$ is hyper-semi-cyclic follows from the definition of T. It remains to show that $(\mathbb{Z}_{v-1},T)$ is a BIBD.

There are $(v-1)\cdot e+2\,\frac{v-1}{3}$ blocks with

$$e = \begin{cases} 2\cdot\#(\text{sextuples}), & \text{if } N_v \text{ is of type a)} \\ \#(\text{sextuples}), & \text{if } N_v \text{ is of type b).} \end{cases}$$

In either case, $e = \frac{v-4}{3}$. Hence, there are

$$(v-1)\cdot e+2\cdot \frac{v-1}{3} = \frac{(v-1)(v-4)}{3} + \frac{2(v-1)}{3}$$
$$= \frac{(v-1)(v-2)}{3}.$$

blocks in T. Next, we will show that $\lambda = 2$. Take any pair of distinct elements $\{k,n\}$ from $\mathbf{Z}_{v-1} \times \mathbf{Z}_{v-1}$. If one of $\{\pm(n-k)\}$ is equal to $\frac{v-1}{3}$ or $\frac{2(v-1)}{3}$, then clearly $\{k,n\} \in T_o$. Hence $\lambda = 2$. If this is not the case, the proof is similar to that of Theorem V.5.

One also checks that each element of $\mathbf{Z}_{v-1}$ appears in exactly v-2 blocks and this completes the proof of the theorem. q.e.d.

Examples:

For v = 16 and a totally proper XIP-neofield of type a) ii) defined as follows:

x	0	1	a	a^2	a^3	a^4	a^5	a^6	a^7	a^8	a^9	a^{10}	a^{11}	a^{12}	a^{13}	a^{14}
$1+x$	1	0	a^4	a^8	a^{14}	a^3	a^{10}	a^{13}	a^9	a^6	a^2	a^5	a^{12}	a	a^7	a^{11}

We then have a type a) XIP-admissible partition of $\mathbf{Z}_{15}$.

$$S(1,4) = \{1,\ 4,\ 3,\ 14,\ 11,\ 12\}$$
$$S(2,8) = \{2,\ 8,\ 6,\ 13,\ 7,\ 9\}$$
$$S(5,10) = \{5,\ 10\}.$$

Hence we have:

$$T(1,4) = \{\{0,1,4\},\ \{1,2,5\},\ \ldots,\ \{14,0,3\}\}$$
$$T(4,3) = \{\{0,4,3\},\ \{1,5,4\},\ \ldots,\ \{14,3,2\}\}$$
$$T(2,8) = \{\{0,2,8\},\ \{1,3,9\},\ \ldots,\ \{14,1,7\}\}$$

$$T(8,6) = \{\{0,8,6\}, \{1,9,7\}, \ldots, \{14,7,5\}\}$$
$$T_o = \{\{0,5,10\}, \{1,6,11\}, \{2,7,12\}, \{3,8,13\}, \{4,9,14\}\}.$$

Let $T = T(1,4) \cup T(4,3) \cup T(2,8) \cup T(8,6) \cup T_o \cup T_o$. One checks that this is an admissible hyper-semi-cyclic BIBD with parameters (15, 70, 14, 3, 2).

We now turn our attention to the cases for a totally proper XIP-neofield of odd order. By Theorem I.52, all the sextuples in an XIP-admissible partition have cardinality 6 except for the singleton $\{0\}$ if $v \equiv 3 \pmod 6$, except for a special triple $\{0,n,-n\}$ if $v \equiv 5 \pmod 6$ and except for a special triple $\{0,n,-n\}$ and the pair $\{\frac{v-1}{3}, \frac{2(v-1)}{3}\}$ if $v \equiv 1 \pmod 6$.

If $v \equiv 3 \pmod 6$, we have e sextuples and one singleton in the XIP-admissible partition of a totally proper XIP-neofield N_v, where

$$e = \begin{cases} \frac{v-3}{6}, & \text{if } N_v \text{ is of type a) ii).} \\ \frac{v-3}{3}, & \text{if } N_v \text{ is of type b) and } t = 0 \text{ in Definition I.58.} \\ \frac{v-3}{3}, & \text{if } N_v \text{ is of type b) and } t \neq 0, \text{ counting twice for the sextuples in } H_t \text{ of Definition I.58.} \end{cases}$$

Hence the total number of blocks arising from these e sextuples is:

$$(v-1)\cdot e = \frac{(v-1)(v-3)}{3}.$$

By adding an ideal element ∞ to $\mathbf{Z}_{v-1}$, let $\mathbf{Z}^+_{v-1} = \mathbf{Z}_{v-1} \cup \{\infty\}$. Let T_o be the subcollection of T:

$$T_o = \{\{0, \ell, \infty\}, \{1, \ell+1, \infty\}, \ldots, \{\ell-1, 2\ell-1, \infty\}\}.$$

Let $T = (\bigcup_{r=1}^{e} T_r) \cup T_o \cup T_o$. If N_v is of type b) and t = 0, each T_r is defined to be T(k,n), associated with S(k,n). If N_v is of type b) and $t \neq 0$, then T_r is defined to be T(k,n), associated with those S(k,n) not in H_t; and to be T(k,n) $\cup$ T(n-k,-k), associated with those S(k,n) in H_t. Finally, if N_v is of type a), T_r = T(k,n) $\cup$ T(n-k,-k) associated with S(k,n) in the type a) admissible partition. Then the total number of blocks in T is:

$$\frac{(v-1)(v-3)}{3} + \frac{2(v-1)}{2} = (v-1) \cdot \frac{v-3+3}{3}$$

$$= \frac{v(v-1)}{3}.$$

By the above arguments and Theorems V.5 and V.13, we have the following:

<u>Theorem V.14</u>: For $v \equiv 3 \pmod 6$ and a totally proper XIP-neofield N_v, $(\mathbf{Z}^+_{v-1}, T)$ defined above is a $(v, \frac{v(v-1)}{3}, v-1, 3, 2)$-BIBD.

<u>Examples</u>:

For v = 9, we have $\mathbf{Z}_8^+ = \{0, 1, 2, \ldots, 7, \infty\}$. A totally proper XIP-neofield of order 9 gives the following type a) XIP-admissible partition of $\mathbf{Z}_8$:

$$S(1,6) = \{1, 6, 5, 7, 2, 3\}$$
$$S(0,0) = \{0\}.$$

Hence we have:

$$T(1,6) = \{\{0,1,6\}, \{1,2,7\}, \ldots, \{7,0,5\}\}$$
$$T(6,5) = \{\{0,6,5\}, \{1,7,6\}, \ldots, \{7,5,4\}\}$$

and

$$T_o = \{\{0,4,\infty\}, \{1,5,\infty\}, \{2,6,\infty\}, \{3,7,\infty\}\}.$$

Let $T = T(1,6) \cup T(6,5) \cup T_o \cup T_o$. One checks that $(\mathbb{Z}_{18}^{+}, T)$ is a BIBD with parameters (9, 24, 8, 3, 2).

For v = 15, we have the following type b) XIP-admissible partition of $\mathbb{Z}_{14}$ from a totally proper XIP-neofield N_{15} (see Sec. III.2):

$$S(1,4) = \{1, 4, 3, 13, 10, 11\}$$
$$S(2,6) = \{2, 6, 4, 12, 8, 10\}$$
$$S(6,1) = \{6, 1, 9, 8, 13, 5\}$$
$$S(5,2) = \{5, 2, 11, 9, 12, 3\}$$
$$S(0,0) = \{0\}.$$

Hence we have:

$$T(1,4) = \{\{0,1,4\}, \{1,2,5\}, \ldots, \{13,0,3\}\}$$
$$T(2,6) = \{\{0,2,6\}, \{1,3,7\}, \ldots, \{13,1,5\}\}$$
$$T(6,1) = \{\{0,6,1\}, \{1,7,2\}, \ldots, \{13,5,0\}\}$$
$$T(5,2) = \{\{0,5,2\}, \{1,6,3\}, \ldots, \{13,4,1\}\}$$
$$T_o = \{\{0,7,\infty\}, \{1,8,\infty\}, \{2,9,\infty\}, \ldots, \{6,13,\infty\}\}.$$

By Theorem V.14, $(\mathbb{Z}_{14}^{+}, T)$ is a (15, 70, 14, 3, 2)-BIBD.

If $v \equiv 5 \pmod 6$, we have e sextuples and one triple $\{0,n,-n\}$ in the XIP-admissible partition of a totally proper XIP-neofield N_v, where

$$e = \begin{cases} \frac{v-5}{6}, & \text{if } N_v \text{ is of type a) ii)} \\ \frac{v-5}{3}, & \text{if } N_v \text{ is of type b) and } t = 0 \text{ in Definition I.58,} \\ \frac{v-5}{3}, & \text{if } N_v \text{ is of type b) and } t \neq 0, \text{ counting twice for the sextuples in } H_t \text{ in Definition I.58} \end{cases}$$

Hence we have the total number of triples arising from these e sextuples is $(v-1)\cdot e = \frac{(v-1)(v-5)}{3}$. Let T_o be the equivalence class:

$$T_o = \{\{0,n,-n\}, \{1,n+1,-n+1\}, \ldots, \{v-2,n+(v-2),-n+(v-2)\}\}.$$

One checks that $|T_o| = v-1$.

Let $T = (\bigcup_{r=1}^{e} T_r) \cup T_o$ and T_r is defined as in the case for $v \equiv 3 \pmod 6$. Then the total number of blocks in T is:

$$\frac{(v-1)(v-5)}{3} + (v-1) = \frac{(v-1)(v-2)}{3}.$$

Then similar to Theorem V.14, we have:

<u>Theorem V.15</u>: For $v \equiv 5 \pmod 6$ and a totally proper XIP-neofield N_v, $(\mathbb{Z}_{v-1}, T)$ defined above is a $(v-1, \frac{(v-1)(v-2)}{3}, v-2, 3, 2)$-BIBD.

Examples:

For v = 23, we have the following type b) XIP-admissible partition of $\mathbb{Z}_{22}$ for a totally proper XIP-neofield N_{23}, constructed in Sec. III.2:

$$S(1,5) = \{1, 5, 4, 21, 17, 18\}$$
$$S(2,7) = \{2, 7, 5, 20, 15, 17\}$$
$$S(3,9) = \{3, 9, 6, 19, 13, 16\}$$
$$S(9,1) = \{9, 1, 14, 13, 21, 8\}$$
$$S(8,2) = \{8, 2, 16, 14, 20, 6\}$$
$$S(7,3) = \{7, 3, 18, 15, 19, 4\}$$
$$S(0,10) = \{0, 10, 12\}.$$

Hence we have:

$$T(1,5) = \{\{0,1,5\}, \{1,2,6\}, \ldots, \{21,0,4\}\}$$
$$T(2,7) = \{\{0,2,7\}, \{1,3,8\}, \ldots, \{21,1,6\}\}$$
$$T(3,9) = \{\{0,3,9\}, \{1,4,10\}, \ldots, \{21,2,8\}\}$$
$$T(9,1) = \{\{0,9,1\}, \{1,10,2\}, \ldots, \{21,8,0\}\}$$
$$T(8,2) = \{\{0,8,2\}, \{1,9,3\}, \ldots, \{21,7,1\}\}$$
$$T(7,3) = \{\{0,7,3\}, \{1,8,4\}, \ldots, \{21,6,2\}\}$$
$$T_o = \{\{0,10,12\}, \{1,11,13\}, \ldots, \{21,9,11\}\}.$$

Let $T = T(1,5) \cup T(2,7) \cup T(3,9) \cup T(9,1) \cup T(8,2) \cup T(7,3) \cup T_o$. One checks that $(\mathbb{Z}_{22}, T)$ is a (22, 154, 21, 3, 2)-BIBD.

If $v \equiv 1 \pmod 6$, we have e sextuples, one triple $\{0,n,-n\}$ and a special pair $\left\{\frac{v-1}{3}, \frac{2(v-1)}{3}\right\}$ in the XIP-admissible partition of a totally proper XIP-neofield N_v, where

$$e = \begin{cases} \frac{v-7}{6}, & \text{if } N_v \text{ is of type a) ii),} \\ \frac{v-7}{3}, & \text{if } N_v \text{ is of type b) and } t = 0 \text{ in Definition I.58} \\ \frac{v-7}{3}, & \text{if } N_v \text{ is of type b) and } t \neq 0 \text{, counting twice for the sextuples in } H_t \text{ of Definition I.58.} \end{cases}$$

Hence the total number of triples arising from these e sextuples is:

$$(v-1)\cdot e = \frac{(v-1)(v-7)}{3}.$$

Let $T_o = T(n,-n)$ and T_o' be the collection of $\frac{v-1}{3}$ triples generated by $\left\{0, \frac{v-1}{3}, \frac{2(v-1)}{3}\right\}$.

Let $T = \left(\bigcup_{r=1}^{e} T_r\right) \cup T_o \cup T_o' \cup T_o'$, where T_r is defined as in the case for $v \equiv 3 \pmod 6$. Then the total number of blocks in T is:

$$\frac{(v-1)(v-7)}{3} + (v-1) + \frac{2(v-1)}{3} = (v-1)\left[\frac{v-7}{3}+1+\frac{2}{3}\right]$$
$$= \frac{(v-1)(v-2)}{3}.$$

Therefore, similar to Theorems V.14 and V.15, we have:

Theorem V.16: For $v \equiv 1 \pmod 6$ and a totally proper XIP-neofield N_v, $(\mathbb{Z}_{v-1}, T)$ as defined above is a $\left(v-1, \frac{(v-1)(v-2)}{3}, v-2, 3, 2\right)$-BIBD.

Example:

For v = 43, we have the following type b) XIP-admissible partition of $\mathbb{Z}_{42}$ for a totally proper

XIP-neofield N_{43} constructed in Sec. III.2:

$$\begin{cases} S(1,4) = \{1, 4, 3, 41, 38, 39\} \\ S(15,32) = \{15, 32, 17, 27, 10, 25\} \\ S(29,18) = \{29, 18, 31, 13, 24, 11\} \end{cases}$$

$$\begin{cases} S(6,1) = \{6, 1, 37, 36, 41, 5\} \\ S(20,29) = \{20, 29, 9, 22, 13, 33\} \\ S(34,15) = \{34, 15, 23, 8, 27, 19\} \end{cases}$$

$$\begin{cases} S(2,6) = \{2, 6, 4, 40, 36, 38\} \\ S(16,34) = \{16, 34, 18, 26, 8, 24\} \\ S(30,20) = \{30, 20, 32, 12, 22, 10\} \end{cases}$$

$$\begin{cases} S(5,2) = \{5, 2, 39, 37, 40, 3\} \\ S(19,30) = \{19, 30, 11, 23, 12, 31\} \\ S(33,16) = \{33, 16, 25, 9, 26, 17\} \end{cases}$$

$$S(0,7) = \{0, 7, 35\}$$

$$S(14,28) = \{14, 28\}.$$

Hence we have:

$$T(1,4) = \{\{0,1,4\}, \{1,2,5\}, \ldots, \{41,0,3\}\}$$

$$T(15,32) = \{\{0,15,32\}, \{1,16,33\}, \ldots, \{41,14,31\}\}$$

$$T(29,18) = \{\{0,29,18\}, \{1,30,19\}, \ldots, \{41,28,17\}\}$$

.

.

.

$$T(33,16) = \{\{0,33,16\}, \{1,34,17\}, \ldots, \{41,32,15\}\}$$

$$T_o = T(7,35) = \{\{0,7,35\}, \{1,8,36\}, \ldots, \{41,6,34\}\}$$

$$T_o' = T(14,28) = \{\{0,14,28\}, \{1,15,29\}, \ldots, \{13,27,41\}\}$$

Let $T = T(1,4) \cup T(15,32) \cup \cdots \cup T(33,16) \cup T_o \cup T_o' \cup T_o''$. Then we have $(\mathbb{Z}_{42}, T)$ is a (42, 574, 41, 3, 2)-BIBD.

Sec. 2. Triple Systems Arising From Even Order LXP-Neofields

In this section, we show that a totally proper LXP-neofield of even order v gives rise to a BIBD with $\lambda = 3$. For an LXP-neofield N_v with v even, let $\mathbb{L}$ be the LXP-admissible partition of $\mathbb{Z}^*_{v-1}$ associated with N_v (see Lemma I.30).

Definition V.17: An LXP-neofield N_v is proper if it is not a CIP (see Chapter I). A proper LXP-neofield is said to be totally proper if for every $\{k,n\} \in \mathbb{L}$, neither $\{n-k,-k\}$, nor $\{-n,k-n\}$ is in $\mathbb{L}$.

For v even, the LXP-admissible partition $\mathbb{L}$ of a totally proper LXP-neofield N_v has the form (Definition I.29):

$$\mathbb{L} = \{\{k_1,n_1\}, \ldots, \{k_{(v-2)/2}, n_{(v-2)/2}\}\}.$$

Let T be defined as follows:

i) $\{0,k,n\} \in T$ if and only if $\{k,n\} \in \mathbb{L}$

ii) $\{i,j,h\} \in T$ if and only if $\{i+1, j+1, h+1\}$ (mod v-1) $\in T$.

Let T(k,n) be the collection of all triples generated by $\{0,k,n\}$. Since (k,n) is admissible in the sense of Definition I.18, T(k,n) is an equivalence class with v-1 triples in each class, i.e., the total number of triples in T is $\frac{(v-1)(v-2)}{2}$.

Each element of $\mathbb{Z}_{v-1}$ appears three times in each equivalent class. Hence

$$r = 3\cdot\frac{(v-2)}{2}.$$

Now, let $\{s,t\}$ be a pair of distinct elements from $\mathbb{Z}_{v-1}$. We assume the following defining relations in the addition table of $\mathbb{Z}_{v-1}$:

$$\begin{cases} a^s+a^t = a^{\ell_1} \\ a^{\ell_2}+a^s = a^t \\ a^s+a^{\ell_3} = a^t \end{cases} \qquad \begin{cases} a^t+a^s = a^{\ell_4} \\ a^{\ell_5}+a^t = a^s \\ a^t+a^{\ell_6} = a^s \end{cases}$$

with $\ell_1 \neq \ell_4$, $\ell_2 \neq \ell_5$, and $\ell_3 \neq \ell_6$ since N_v is totally proper. We then have:

$$\begin{aligned} a^s+a^t = a^{\ell_1} &\Rightarrow 1+a^{t-s} = a^{\ell_1-s} \\ &\Rightarrow 1+a^{\ell_1-s} = a^{t-s} \text{ (by a) of Lemma I.16)} \\ &\Rightarrow a^s+a^{\ell_1} = a^t. \end{aligned}$$

Hence we have $\ell_1 \equiv \ell_3$ (mod v-1). Similarly, we have $\ell_2 \equiv \ell_5$ (mod v-1) and $\ell_4 \equiv \ell_6$ (mod v-1). Then

$$\{s,t\} \subset \{s,t,\ell_1\} = \{s,\ell_1,t\} \in T(t-s,\ell_1-s)$$
$$\{s,t\} \subset \{\ell_2,s,t\} = \{\ell_2,t,s\} \in T(s-\ell_2,t-\ell_2)$$

and

$$\{s,t\} \subset \{t,s,\ell_4\} = \{t,\ell_4,s\} \in T(s-t,\ell_4-t)$$

Therefore $\lambda = 3$. Summarizing the above arguments, we have the following theorem:

Theorem V.18: For v even and a totally proper LXP-neofield N_v, the triple system $(\mathbb{Z}_{v-1},T)$ derived above is a $(v-1, \frac{(v-1)(v-2)}{2}, \frac{3(v-2)}{2}, 3, 3)$-BIBD.

Examples:

For v = 8. We have the following totally proper LXP-neofield N_8.

x	0	1	a	a^2	a^3	a^4	a^5	a^6
$1+x$	1	0	a^6	a^5	a^4	a^3	a^2	a .

Hence we have the LXP-admissible partition:

$$L = \{\{1,6\}, \{2,5\}, \{3,4\}\}.$$

Let

$$T(1,6) = \{\{0,1,6\}, \{1,2,0\}, \ldots, \{6,0,5\}\}$$
$$T(2,5) = \{\{0,2,5\}, \{1,3,6\}, \ldots, \{6,1,4\}\}$$
$$T(3,4) = \{\{0,3,4\}, \{1,4,5\}, \ldots, \{6,2,3\}\}$$

and

$$T = T(1,6) \cup T(2,5) \cup T(3,4).$$

One checks that $(\mathbb{Z}_7,T)$ is a (7, 21, 9, 3, 3)-BIBD.

For v = 10. We have the following LXP-admissible partition of a totally proper LXP-neofield N_{10}.

$$L = \{\{1,2\}, \{3,6\}, \{4,8\}, \{5,7\}\}.$$

Then

$$T(1,2) = \{\{0,1,2\}, \{1,2,3\}, \ldots, \{8,0,1\}\}$$
$$T(3,6) = \{\{0,3,6\}, \{1,4,7\}, \ldots, \{8,2,5\}\}$$
$$T(4,8) = \{\{0,4,8\}, \{1,5,0\}, \ldots, \{8,3,7\}\}$$
$$T(5,7) = \{\{0,5,7\}, \{1,6,8\}, \ldots, \{8,4,6\}\} .$$

Let $T = T(1,2) \cup T(3,6) \cup T(4,8) \cup T(5,7)$. Then $(\mathbb{Z}_9, T)$ is a (9, 36, 12, 3, 3)-BIBD.

Sec. 3. Triple Systems Arising From Even Order XMP-Neofields

In this section, we show that a totally proper XMP-neofield of even order v gives rise to a BIBD with $\lambda = 6$. For an XMP-neofield N_v with v even, let X be the XMP-admissible partition of $\mathbb{Z}_{v-1}^*$ associated with N_v (see Lemma I.22).

Definition V.19: An XMP-neofield N_v is proper if it is not an XIP, not an LXP, not a CMP and not an RXP (see Chapter I). A proper XMP-neofield is said to be totally proper if for every $(k,n) \in X$, neither $(n-k,-k)$, nor $(-n,k-n)$ is in X.

The XMP-admissible partition X of a totally proper XMP-neofield N_v has the form (definition I.21).

$$X = \{(k_1,n_1), (k_2,n_2), \ldots, (k_{v-2},n_{v-2})\}.$$

Let T be defined as follows:

$$\begin{cases} \text{i)} & \{0,k,n\} \in T \text{ if } (k,n) \in X \\ \text{ii)} & \{i,j,h\} \in T \text{ if } \{i+1,\ j+1,\ h+1\} (\text{mod } v-1) \in T. \end{cases}$$

Let T(k,n) be the collection of all triples generated by $\{0,k,n\}$. Since (k,n) is admissible in the sense of Definition I.18, T(k,n) is an equivalence class with v-1 triples. Hence T is divided into v-2 equivalent classes with v-1 triples in each class, i.e., the total number of triples in T is (v-1)(v-2).

Each element of $\mathbb{Z}_{v-1}$ appears three times in each equivalent class. Hence r = 3(v-2). Now, let $\{s,t\}$ be a pair of distinct element from $\mathbb{Z}_{v-1}$. We assume the following defining relations in the addition table of N_v:

$$\begin{cases} a^s+a^t = a^{\ell_1} \\ a^{\ell_2}+a^s = a^t \\ a^s+a^{\ell_3} = a^t \end{cases} \qquad \begin{cases} a^t+a^s = a^{\ell_4} \\ a^{\ell_5}+a^t = a^s \\ a^t+a^{\ell_6} = a^s, \end{cases}$$

with $\ell_1 \neq \ell_4$, $\ell_2 \neq \ell_5$, and $\ell_3 \neq \ell_6$ since N_v is not CMP. Moreover, we claim that $|\{\ell_1,\ell_2,\ell_3,\ell_4,\ell_5,\ell_6\}| = 6$. We only show that $\ell_1 \neq \ell_2$. The remaining cases are similar. If $\ell_1 = \ell_2$, then:

$$(a^s+a^t)+a^s = a^t.$$

This is impossible, since N_v is totally proper and hence it doesn't have XIP. We have:

$$\{s,t\} \subset \{s,t,\ell_1\} \in T(t-s,\ell_1-s)$$
$$\{s,t\} \subset \{\ell_2,s,t\} \in T(s-\ell_2,t-\ell_2)$$
$$\{s,t\} \subset \{s,\ell_3,t\} \in T(\ell_3-s,t-s)$$

and

$$\{s,t\}\subset\{t,s,\ell_4\}\in T(s-t,\ell_4-t)$$
$$\{s,t\}\subset\{\ell_5,t,s\}\in T(t-\ell_5,s-\ell_5)$$
$$\{s,t\}\subset\{t,\ell_6,s\}\in T(\ell_6-t,s-t).$$

Therefore $\lambda = 6$. Summarizing the above arguments, we have the following:

<u>Theorem V.20</u>: For v even and a totally proper XMP-neofield N_v, the triple system $(\mathbb{Z}_{v-1},T)$ derived above is a (v-1, (v-1)(v-2), 3(v-2), 3, 6)-BIBD.

<u>Example</u>:

For v = 10. We have the following XMP-admissible partition of a totally proper XMP-neofield N_{10}:

$$X = \{(1,8), (2,7), (3,5), (4,3), (5,6), (6,1), (7,4), (8,2)\}$$

Then

$$T(1,8) = \{\{0,1,8\}, \{1,2,0\}, \ldots, \{8,0,7\}\}$$
$$T(2,7) = \{\{0,2,7\}, \{1,3,8\}, \ldots, \{8,1,6\}\}$$
$$T(3,5) = \{\{0,3,5\}, \{1,4,6\}, \ldots, \{8,2,4\}\}$$
$$T(4,3) = \{\{0,4,3\}, \{1,5,4\}, \ldots, \{8,3,2\}\}$$
$$T(5,6) = \{\{0,5,6\}, \{1,6,7\}, \ldots, \{8,4,5\}\}$$
$$T(6,1) = \{\{0,6.1\}, \{1,7,2\}, \ldots, \{8,5,0\}\}$$
$$T(7,4) = \{\{0,7,4\}, \{1,8,5\}, \ldots, \{8,6,3\}\}$$
$$T(8,2) = \{\{0,8,2\}, \{1,0,3\}, \ldots, \{8,7,1\}\}.$$

Let $T = T(1,8)\cup T(2,7)\cup T(3,5)\cup T(4,3)\cup T(5,6)\cup T(6,1)\cup T(7,4)$ $\cup T(8,2)$.

Hence $(\mathbb{Z}_9, T)$ is a (9, 72, 24, 3, 6)-BIBD.

Sec. 4. A Reconstruction Theorem

In this section, it is shown that if $v \equiv 2 \pmod 6$, a totally proper XIP-neofield N_v can be reconstructed from an admissible semi-cyclic BIBD with parameters $(v-1, \frac{(v-1)(v-2)}{3}, v-2, 3, 2)$.

For $m \equiv 1 \pmod 6$, let $(\mathbb{Z}_m, \mathcal{B})$ be an admissible semi-cyclic $(m, \frac{m(m-1)}{3}, m-1, 3, 2)$-BIBD defined as in Definition V.8. Let

$$\mathcal{B} = \mathcal{B}_1 \cup \mathcal{B}_2 \cup \ldots \cup \mathcal{B}_e,$$

where $e = \frac{m-1}{3}$.

Lemma V.21: If $\{0,k,n\} \in \mathcal{B}_i$ for some i, then $\{-k,0,n-k\}$ and $\{-n,k-n,0\}$ are also in $\mathcal{B}_i$.

Proof: It follows from Definition V.4. q.e.d.

Lemma V.22: For any $x \in \mathbb{Z}_m - \{0\}$, there exist $y_1, y_2 \in \mathbb{Z}_m - \{0\}$, $y_1 \neq y_2$ such that $\{0,x,y_1\}$ and $\{0,y_2,x\}$ are all in $\mathcal{B}$.

Proof: Since $(\mathbb{Z}_m, \mathcal{B})$ is a BIBD with $\lambda = 2$, $x \in \mathbb{Z}_m - \{0\}$ implies that $\{0,x\}$ appears as subset of two of the blocks in $\mathcal{B}$. Hence the lemma follows. q.e.d.

If $\{0,k,n\} \in \mathcal{B}_i$, let $S_i = \{k,n,n-k,-k,-n,k-n\}$. We also adopt the terminologies of odd-parity and even-parity elements defined in Sec. I.3.

Let π be a permutation of $\mathbb{Z}_m - \{0\}$ with the following condition:

$$\begin{cases} \text{a) for } S_i = \{k,n,n-k,-k,-n,k-n\} \text{ derived from } \{0,k,n\} \in \beta_i, \text{ let } \pi(k) = n, \pi(n-k) = -k, \text{ and } \pi(-n) = k-n; \\ \text{b) define } S_1, S_2, \ldots, S_e \text{ successively such that each element } x \text{ in } \mathbb{Z}_m-\{0\} \text{ appears in two } S_i\text{'s and } x \text{ is of odd-parity in one } S_i \text{ and of even-parity in the other } S_i \text{ (Lemma V.22).} \end{cases}$$

Hence we have the following reconstruction theorem:

<u>Theorem V.23</u>: For $m \equiv 1 \pmod 6$, let $G = \{1, a, a^2, \ldots, a^{m-1}\}$ be the cyclic group of order m. Let $S = G \cup \{0\}$, with multiplication in S the extension of multiplication in G given by

$$0 \cdot g = g \cdot 0 = 0, \forall g \in S.$$

We define an addition in S by:

$$\begin{cases} 0+x = x+0 = x, \forall x \in S \\ 1+1 = 0 \\ 1+a^k = a^n, \text{ if } \pi(k) = n \\ a^r+a^s = a^r(1+a^{s-r}) \text{ for } r, s \neq 0. \end{cases}$$

This addition is well-defined and $N_v = \langle S,+,\cdot \rangle$ is a totally proper XIP-neofield of order $v = m+1$.

<u>Proof</u>: By condition b) of the permutation π,

$$\bigcup_{i=1}^{e} \{\text{odd-parity elements in } S_i\} = \mathbb{Z}_m-\{0\},$$

since each x of $\mathbb{Z}_m - \{0\}$ is of odd-parity in one S_i. On the other hand, for each sextuple

$$S_i = \{k,\ n,\ n-k,\ -k,\ -n,\ k-n\},$$

we have three mappings $\pi(k) = n$, $\pi(n-k) = -k$, and $\pi(-n) = k-n$. Hence

$$\{\pi(x)-x \mid x=k,\ n-k,\ -n\} = \{n-k,\ -n,\ k\}$$
$$= \{\text{odd-parity elements in } S_i\}.$$

Then

$$\bigcup_{x\in\mathbb{Z}_m^*} \{\pi(x)-x\} = \bigcup_{i=1}^{e} \{\text{odd-parity elements in } S_i\}$$
$$= \mathbb{Z}_m^*$$

It follows that π is a permutation with $\pi(x)-x$ all distinct. Hence the presentation function T such that $T(a^k) = a^n$ if $\pi(k) = n$, defines a cyclic neofield of even order, after applying the distributive law.

The fact that N_v has XIP follows from the definition of π, since each of the following three statements implies the remaining two:

$$\begin{cases} \text{i)}\ \pi(k) = n \\ \text{ii)}\ \pi(n-k) = -k \\ \text{iii)}\ \pi(-n) = k-n. \end{cases}$$

Moreover, N_v is a totally proper XIP-neofield. Otherwise, we would get less than $\frac{m-1}{3}$ equivalence classes. This completes the proof. q.e.d.

Example:

For $m = 7 \equiv 1 \pmod 6$, let $(\mathbb{Z}_7, \mathcal{B})$ be an admissible semi-cyclic BIBD of parameters (7, 14, 6, 3, 2) defined as follows:

$$\mathcal{B} = \mathcal{B}_1 \cup \mathcal{B}_2, \text{ where}$$

$$\mathcal{B}_1 = \{\{0,1,5\}, \{1,2,6\}, \ldots, \{6,0,4\}\}$$
$$\mathcal{B}_2 = \{\{0,3,1\}, \{1,4,2\}, \ldots, \{6,2,0\}\}.$$

Let π be a permutation of $\mathbb{Z}_7 - \{0\}$ defined as:

x	1	2	3	4	5	6
$\pi(x)$	5	3	1	6	4	2

Hence we have reconstructed the following totally proper XIP-neofield of order 8:

x	0	1	a	a^2	a^3	a^4	a^5	a^6
$1+x$	1	0	a^5	a^3	a	a^6	a^4	a^2

by Theorem V.23.

Sec. 5. The Interrelation of Cyclic Neofields in Terms of Designs

We conclude this chapter by studying the interrelation of cyclic neofields with different algebraic properties in terms of triple systems.

Johnsen[14] has shown that the existence of a CIP-neofield of even order v is equivalent to the existence of a cyclic Steiner triple system of order v-1, i.e., a (v-1, 3, 1) symmetric BIBD.

In Sec. V.1, we have shown that a totally proper XIP-neofield of even order v gives rise to a (v-1, $\frac{(v-1)(v-2)}{3}$, v-2, 3, 2)-BIBD. In Sec. V.2, we have shown (Theorem V.18) that a totally proper LXP-neofield of even order v gives rise to a (v-1, $\frac{(v-1)(v-2)}{2}$, $\frac{3(v-2)}{2}$, 3, 3)-BIBD.

In Sec. V.3, we have shown (Theorem V.20) that a totally proper XMP-neofield of even order v gives rise to a (v-1, (v-1)(v-2), 3(v-2), 3, 6)-BIBD. Hence the lattice relation of different property cyclic neofields analogous to Galois Theory makes sense. The stronger the algebraic property a cyclic neofield has, the smaller the number λ of the BIBD arising from this cyclic neofield.

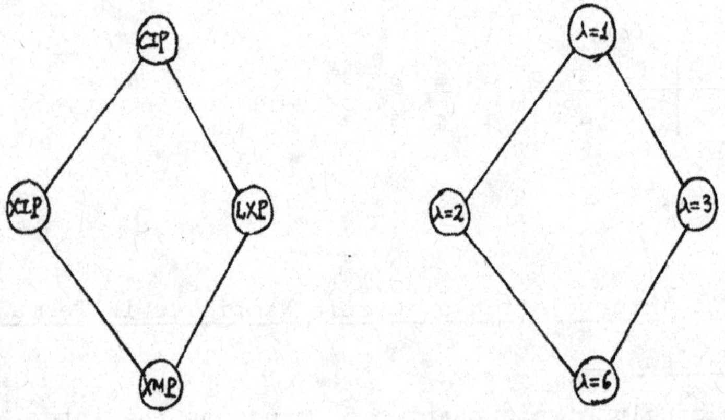

Figure 3

CHAPTER VI

CYCLIC NEOFIELDS AND PERMUTATION MATRICES

Sec. 1. Definitions and Relations

In this section, we define I-matrices, N-permutations, and N-graphs and study the relation between them and cyclic neofields. For the terminology of directed graphs, see [6]. For any other terminologies on combinatories, see [5].

Note that we let $\mathbb{Z}_n^*$ be the set $\mathbb{Z}_n - \{0\}$ if n is odd, and the set $\mathbb{Z}_n - \{\frac{n}{2}\}$ if n is even.

Definition VI.1: A permutation π on $\mathbb{Z}_n^*$ is called an N-permutation if $\{\pi(x)-x \mid x \in \mathbb{Z}_n^*\} = \mathbb{Z}_n^*$.

Examples:

We give 6 examples to illustrate N-permutations on $\mathbb{Z}_n^*$ with different n.

n = 5,

x	1	2	3	4
$\pi(x)$	2	4	1	3
$\pi(x)-x$	1	2	3	4

n = 6,

x	0	1	2	4	5
$\pi(x)$	4	2	1	0	5
$\pi(x)-x$	4	1	5	2	0

n = 7,

x	1	2	3	4	5	6
π(x)	3	5	2	1	6	4
π(x)-x	2	3	6	4	1	5

n = 8,

x	0	1	2	3	5	6	7
π(x)	5	7	1	3	6	0	2
π(x)-x	5	6	7	0	1	2	3

n = 9,

x	1	2	3	4	5	6	7	8
π(x)	4	3	7	2	1	8	6	5
π(x)-x	3	1	4	7	5	2	8	6

n = 10,

x	0	1	2	3	4	6	7	8	9
π(x)	4	3	8	2	1	6	0	9	7
π(x)-x	4	2	6	9	7	0	3	1	8

Let $A_m = [a_{ij}]$ be a permutation matrix of order m. Here the rows and columns of a matrix of order m are labeled in the order: 1, 2, 3, ..., m if m is even, and: 0, 1, 2, 3, ..., $\frac{m-1}{2}$, $\frac{m+3}{2}$, ..., m, if m is odd. So it is understood that the entry a_{ij} is in the i-th row and the j-th column. Throughout this chapter, we are only interested in the permutation matrix $A_m = [a_{ij}]$ with $a_{ii} = 0$ if m is even and $a_{ij} = 0$ for $j-i \equiv \frac{m+1}{2} \pmod{m+1}$ if m is odd.

<u>Definition VI.2</u>: For a permutation matrix $A_m = \{a_{ij}\}$, entries a_{ij} and a_{hk} are said to be on the same diagonal of A_m (called these diagonals <u>special diagonals</u>) if $j-i \equiv k-h \pmod{m+1}$.

For m even, by excluding the main diagonal with entries $a_{ii} = 0$, i = 1, 2, ..., m, we have m such special diagonals with m-1 entries each. For m odd, by excluding entries $a_{ij} = 0$ with $j-i \equiv \frac{m+1}{2} \pmod{m+1}$ (there are m-1 of these entries), we have m such special diagonals all of them having m entries a_{ii}, $i = 0, 1, 2, \ldots, \frac{m-1}{2}, \frac{m+3}{2}, \ldots, m$.

Definition VI.3: An I-matrix A_m of order m is a permutation matrix of order m with $a_{ii} = 0$ if m is even and $a_{ij} = 0$ for $j-i \equiv \frac{m+1}{2} \pmod{m+1}$ if m is odd, and the property that there exists only one 1 on each of the m special diagonals.

Examples: I-matrices A_m of different orders

m even:

$m = 2$,

$$\begin{bmatrix} 0 & 1 \\ 1 & 0 \end{bmatrix}$$

$m = 4$,

$$\begin{bmatrix} 0 & 1 & 0 & 0 \\ 0 & 0 & 0 & 1 \\ 1 & 0 & 0 & 0 \\ 0 & 0 & 1 & 0 \end{bmatrix} \quad \begin{bmatrix} 0 & 0 & 1 & 0 \\ 1 & 0 & 0 & 0 \\ 0 & 0 & 0 & 1 \\ 0 & 1 & 0 & 0 \end{bmatrix} \quad \begin{bmatrix} 0 & 0 & 0 & 1 \\ 0 & 0 & 1 & 0 \\ 0 & 1 & 0 & 0 \\ 1 & 0 & 0 & 0 \end{bmatrix}$$

$m = 6$,

$$\begin{bmatrix} 0 & 0 & 0 & 0 & 1 & 0 \\ 1 & 0 & 0 & 0 & 0 & 0 \\ 0 & 0 & 0 & 1 & 0 & 0 \\ 0 & 0 & 0 & 0 & 0 & 1 \\ 0 & 0 & 1 & 0 & 0 & 0 \\ 0 & 1 & 0 & 0 & 0 & 0 \end{bmatrix} \quad \begin{bmatrix} 0 & 0 & 0 & 0 & 0 & 1 \\ 0 & 0 & 0 & 0 & 1 & 0 \\ 0 & 0 & 0 & 1 & 0 & 0 \\ 0 & 0 & 1 & 0 & 0 & 0 \\ 0 & 1 & 0 & 0 & 0 & 0 \\ 1 & 0 & 0 & 0 & 0 & 0 \end{bmatrix} \quad \ldots$$

m odd:

m = 3,

$$\begin{array}{c|ccc} & 0 & 1 & 3 \\ \hline 0 & 0 & 0 & 1 \\ 1 & 0 & 1 & 0 \\ 3 & 1 & 0 & 0 \end{array} \qquad \begin{bmatrix} 0 & 1 & 0 \\ 1 & 0 & 0 \\ 0 & 0 & 1 \end{bmatrix}$$

m = 5,

$$\begin{array}{c|ccccc} & 0 & 1 & 2 & 4 & 5 \\ \hline 0 & 0 & 0 & 0 & 1 & 0 \\ 1 & 0 & 0 & 1 & 0 & 0 \\ 2 & 0 & 1 & 0 & 0 & 0 \\ 4 & 1 & 0 & 0 & 0 & 0 \\ 5 & 0 & 0 & 0 & 0 & 1 \end{array} \qquad \begin{bmatrix} 0 & 0 & 1 & 0 & 0 \\ 0 & 1 & 0 & 0 & 0 \\ 1 & 0 & 0 & 0 & 0 \\ 0 & 0 & 0 & 0 & 1 \\ 0 & 0 & 0 & 1 & 0 \end{bmatrix} \qquad \ldots$$

Now we turn our attention to directed graphs. Let G_m be a directed graph with m points. If m is even, we label the points of G_m as: 1, 2, 3, ..., m. If m is odd, we label the points of G_m as: 0, 1, 2, ..., $\frac{m-1}{2}$, $\frac{m+3}{2}$, ..., m.

<u>Definition VI.4</u>: Let G_m be a directed graph labelled as above. Then G_m is called an <u>N-graph of order m</u> if (1) every point of G_m has in-degree 1 and out-degree 1, except there is a loop at a certain point when m is odd, (2) we have

$$\left\{ v-u \ (\text{mod } m+1) \;\middle|\; (u,v) \text{ is an edge of } G_m \right\} = \mathbb{Z}^*_{m+1}$$

Condition (1) above indicates that we have a permutation π of the set $\mathbb{Z}^*_{m+1}$. Condition (2) simply says that $\pi(x)-x$ are all distinct in $\mathbb{Z}^*_{m+1}$. Also note that an N-permutation on $\mathbb{Z}^*_n$ with n odd is equivalent to an A-permutation π, defined on $\mathbb{Z}_n$ in Sec. IV.3, with $\pi(0) = 0$.

<u>Examples</u>: N-graphs of order m.

m = 2,

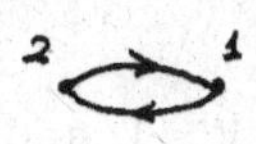

m = 4,

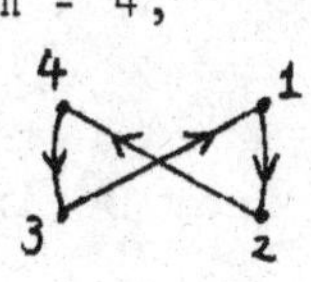

m = 6,

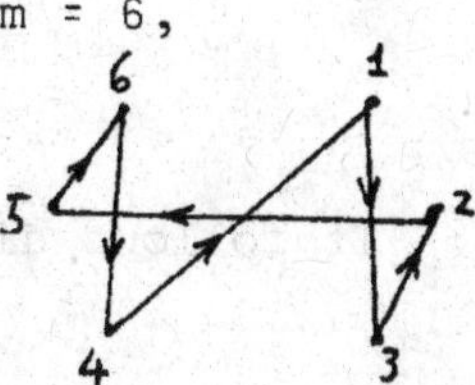

m = 8,

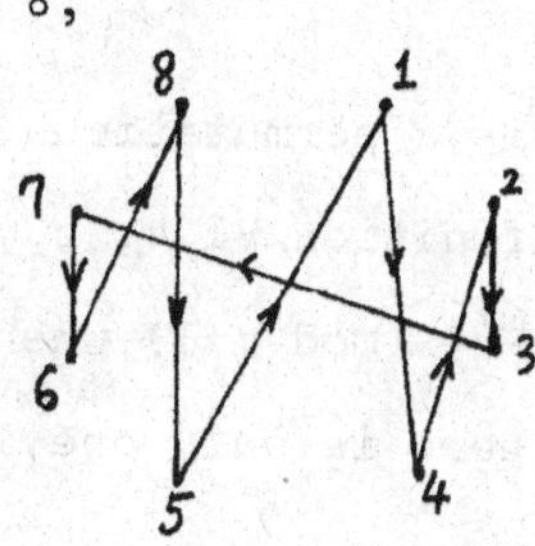

. . .

m = 3,

0
3
1

m = 5,

m = 7,

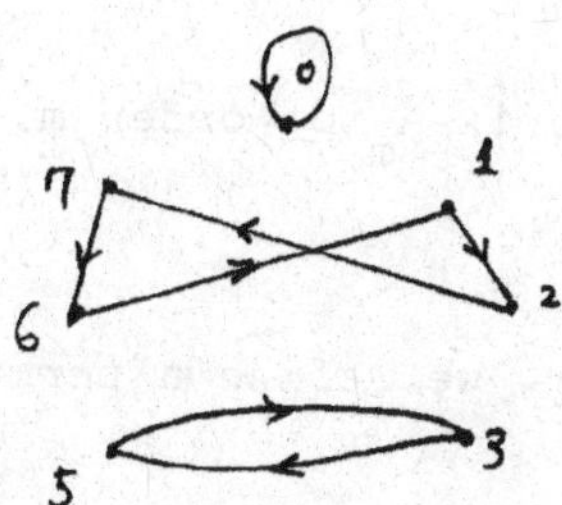

. . .

<u>Lemma VI.5</u>: The existence of an N-permutation π on $\mathbb{Z}_n^*$ is equivalent to the existence of an N-graph G_{n-1} of order n-1.

<u>Proof</u>: It follows directly from Definitions VI.1 and VI.4.

q.e.d.

<u>Lemma VI.6</u>: Let G be an N-graph of order m and M(G) be the adjacency matrix of G. Then M(G) is an I-matrix of order m.

<u>Proof</u>: By (1) of Definition VI.4, M(G) is a permutation matrix. Let $M(G) = [a_{ij}]$. By (2) of Definition VI.4, $a_{ii} = 0$, when m is even and $a_{ij} = 0$ for $j-i \equiv \frac{m+1}{2} \pmod{m+1}$ when m is odd. It remains to be shown that there is only one 1 on each of the special diagonals of M(G).

We only show the case for m even. For the case with m odd, the proof is similar. So let m be even, we assume there are two 1's on some of those m special diagonals, say $m_{ij} = m_{hk} = 1$, where $j-i \equiv k-h \pmod{m+1}$. Hence there are edges (i,j) from i to j and (h,k) from h to k in G with $j-i \equiv k-h \pmod{m+1}$. This contradicts condition (2) of Definition VI.4.

q.e.d.

<u>Lemma VI.7</u>: If there exists an I-matrix A_m of order m, there exists an N-permutation on $\mathbb{Z}_{m+1}^*$.

<u>Proof</u>: For an I-matrix A_m of order m, we define a permutation π on $\mathbb{Z}_{m+1}^*$ as follows:

$$\pi(x) = y \Leftrightarrow \quad (x,y) \text{ entry of } A_m = 1$$

We assume $\pi(x_1)-x_1 = \pi(x_2)-x_2$ for some $x_1 \neq x_2$ in $\mathbb{Z}_m^*$. By Definition VI.2, these two entries $(x_1,\pi(x_1))$ and $(x_2,\pi(x_2))$ are on the same special diagonal with $(x_1,\pi(x_1)) = (x_2, \pi(x_2)) = 1$. This contradicts Definition VI.3. Therefore $\pi(x)-x$ are all distinct in $\mathbb{Z}_m^*$ when x runs over the whole set $\mathbb{Z}_m^*$. <u>q.e.d.</u>

<u>Theorem VI.8</u>: The following statements are equivalent to each other:

(1) there exists a cyclic neofield N_v of order v,
(2) there exists an I-matrix A_{v-2} of order v-2,
(3) there exists an N-permutation π of $\mathbb{Z}_{v-1}^*$,
(4) there exists an N-graph G_{v-2} of order v-2.

<u>Proof</u>: By Lemma VI.5, we have that (3) and (4) imply each other. By Lemma VI.6, (4) implies (2). By Lemma VI.7, (2) implies (3). It is then sufficient to show that (1) and (3) imply each other.

Assume (1) is true. Let N_v be a cyclic neofield of order v. By Definition I.22, there is an XMP-admissible partition XM. Let π be a permutation on $\mathbb{Z}_{v-1}^*$ defined as $\pi(x) = y$ if the ordered pair (x,y) is in XM. By Definition I.21, $\pi(x)-x$ runs over all elements in $\mathbb{Z}_{v-1}^*$. This implies (3).

Next assume (3) is true. Let π be an N-permutation of $\mathbb{Z}_{v-1}^*$. Define a partition XM of $\mathbb{Z}_{v-1}^*$ to be that $(k,n) \in XM$ if $\pi(k) = n$. This gives an XMP-admissible partition of $\mathbb{Z}_{v-1}^*$. Hence by Theorem I.23, we can construct a cyclic neofield. This completes the proof of the theorem.

q.e.d.

Sec. 2. Decomposition Theorem

Let J_m^* be the matrix such that $a_{ii} = 0$, $a_{ij} = 1$, $\forall i \neq j$. By Konig's Theorem([17], p.239), J_m^* can be decomposed into a sum of m-1 permutation matrices of order m in the following way:

$$J_m^* = \sum_{i=1}^{m-1} P_i,$$

where P_i denotes a permutation matrix which has its entry 1 and row 1 on column i, and the rows and columns are labeled in order: 1, 2, 3, ..., m. This is called a Konig decomposition of J_m^* by Johnsen and Storer [10]. We now give a result analogous to this, but the special diagonals of the permutation matrices are also involved.

Lemma VI.9: Let P = 2n+1 be an odd prime. Let π_i be a mapping from $\mathbf{Z}_p^*$ to $\mathbf{Z}_p^*$ defined by $\pi_i(x) = i \cdot x$, $x \in \mathbf{Z}_p^*$, where i = 2, 3, 4, ..., 2n and $\mathbf{Z}_p^* = \mathbf{Z}_p - \{0\}$. Then π_i is a derangement on $\mathbf{Z}_p^*$ for all i.

Proof: If $\pi_i(j) \equiv j \pmod p$ for some j, then we have $i \cdot j \equiv j \pmod p$ for some j. Hence $(i-1) \cdot j \equiv 0 \pmod p$. But this is impossible since $i \in \{2, 3, 4, \ldots, 2n\}$, $j < P$ and p is a prime number. Therefore π_i moves every number in $\mathbf{Z}_p^*$. To show that π_i is a permutation, we assume $\pi_i(j_1) = \pi_i(j_2)$. Then $i \cdot j_1 \equiv i \cdot j_2 \pmod p$. Hence $i \cdot (j_1 - j_2) \equiv 0 \pmod p$ and then $j_1 - j_2 \equiv 0 \pmod{p}$ since $(i,p) = 1$. Thus $j_1 \equiv j_2 \pmod{p}$. This completes the proof. q.e.d.

Theorem VI.10: If $p = 2n+1$ is an odd prime, then J^*_{2n} can be decomposed into the sum of $2n-1$ I-matrices.

Proof: We define π_i as in Lemma VI.9. By the same lemma, π_i is both a derangement and a permutation on Z^*_p for all $i = 2, 3, \ldots, 2n$. For each i, let $A^{(i)}_{2n} = [a_{hk}]$ be the matrix such that $a_{hk} = 1$ if $\pi_i(h) = k$ and $a_{hk} = 0$ otherwise. Since π_i is a permutation, $A^{(i)}_{2n}$ is a permutation matrix. Since π_i is a derangement, $A^{(i)}_{2n}$ has the property that $a_{hh} = 0$, $h = 1, 2, \ldots, 2n$. We will show that $A^{(i)}_{2n}$ is an I-matrix of order $2n$, for $i = 2, 3, \ldots, 2n$.

Assume there are two 1's on the same special diagonal of $A^{(i)}_{2n}$. Say they are on j_1th and j_2th row respectively. Then they must be on the entries (j_1, ij_1) and (j_2, ij_2) of $A^{(i)}_{2n}$. Therefore we have $i \cdot j_1 - j_1 \equiv i \cdot j_2 - j_2 \pmod{p}$. Hence $(i-1) \cdot j_1 \equiv (i-1) \cdot j_2 \pmod{p}$, i.e., $(i-1) \cdot (j_1 - j_2) \equiv 0 \pmod{p}$ and then $j_1 \equiv j_2 \pmod{p}$. Thus $A^{(i)}_{2n}$ is an I-matrix of order $2n$.

We will show that $A^{(i_1)}_{2n}$ and $A^{(i_2)}_{2n}$ are disjoint for $i_1 \neq i_2$ and $i_1, i_2 \in \{2, 3, 4, \ldots, 2n\}$. Assume $a_{hk}{}^{(i_1)} = a_{hk}{}^{(i_2)}$, where $a_{hk}{}^{(i_1)} \in A^{(i_1)}_{2n}$ and $a_{hk}{}^{(i_2)} \in A^{(i_2)}_{2n}$. Hence $\pi_{i_1}(h) = k$, $\pi_{i_2}(h) = k$ and then $i_1 \cdot h \equiv k \equiv i_2 \cdot h \pmod{p}$. Thus $(i_1 - i_2) \cdot h \equiv 0 \pmod{p}$, i.e., $i_1 \equiv i_2 \pmod{p}$, a contradiction.

There are $(2n \times 2n) - 2n$ 1's in J^*_{2n} and each 1 can only appear in exactly one of $A_{2n}^{(i)}$, i = 2, 3, 4, ..., 2n. We then have

$$\begin{aligned} &\# \text{ of non-zero entries in } J_{2n}^{*} \\ &= (2n \times 2n)-2n \\ &= (2n-1)\cdot 2n \\ &= (2n-1)\cdot \# \text{ of non-zero entries in each } A_{2n}^{(i)} \\ &= \# \text{ of non-zero entries in } \sum_{i=2}^{2n} A_{2n}^{(i)}. \end{aligned}$$

Hence J_{2n}^{*} is decomposed into the sum of 2n-1 I-matrices $A_{2n}^{(i)}$. <u>q.e.d.</u>

This decomposition is called a <u>3-dimension Konig decomposition</u> of J_{2n}^{*}. This result is then used to study factorization of complete directed graphs.

<u>Corollary VI.11</u>: If p = 2n+1 is an odd prime, then the complete directed graph D_{2n} has a factorization into 2n-1 N-graphs.

<u>Sec. 3. Structure on I-Matrices</u>

In this section, we introduce a structure on the set of I-matrices of given order <u>via</u> a set of transformations $\{t, dr, dc, \text{and } \Delta\}$. We only study the case when the order is even, the odd order case being similar.

For convenience, let F_{2m} be the family of all I-matrices of order 2m and let $I_{k,n}$ be the (k,n) entry of the I-matrix I, where subindices are in $\mathbb{Z}_{2m+1}$ if the order of I is 2m. We will introduce a set of transformations $\{t, dr,$

dc, and $\Delta\}$ on the family F_{2m}.

<u>Definition VI.12</u>: Let I be an I-matrix of order 2m. Let the transformations i, t, dr, dc, and Δ be defined as follows:

$$\begin{cases} (1) & [t(I)]_{n,k} = I_{k,n} \\ (2) & [dr(I)]_{k-n,-n} = I_{k,n} \\ (3) & [dc(I)]_{-k,n-k} = I_{k,n} \\ (4) & [\Delta(I)]_{n-k,-k} = I_{k,n} \\ (5) & [i(I)]_{k,n} = I_{k,n} \end{cases}$$

<u>Lemma VI.13</u>: The transformations t, dr, dc, Δ are well-defined, i.e., t(I), dr(I), dc(I) and (I) are all I-matrices if I is an I-matrix.

<u>Proof</u>: It is obvious that t is well-defined. The proofs of the remaining cases are similar to each other. Hence we only show the case of dr.

Let I_{2m} be an I-matrix of order 2m. Let $dr(I_{2m}) = M$. First, we show that M is a permutation matrix. Assume $M_{i,j_1} = M_{i,j_2} = 1$. Then by (2) of Definition VI.12, $I_{i-j_1,-j_1} = I_{i-j_2,-j2} = 1$ in I_{2m}. This is true only if $-j_1 \equiv -j_2 \pmod{2m+1}$, i.e., $j_1 \equiv j_2 \pmod{2m+1}$. Hence there is only one 1 in each row of M. Similarly, there is only one 1 in each column of M. Next, we show that $M_{i,i} = 0$ for all i from 1 to 2m. Assume this is not the case, ie., $M_{i,i} = 1$ for some i. Then by (2) of Definition VI.12, $I_{i-i,-i} = I_{0,-i} = 1$, which is absurd.

It remains to show that in $dr(I_{2m}) = M$, there is only one 1 on each of the 2m-1 special diagonals. Assume $M_{i,j} = M_{h,k} = 1$ with $j-i \equiv k-h \pmod{2m+1}$. Then by (2) of Definition VI.12, $I_{i-j,-j} = I_{h-k,-k} = 1$ and keep in mind that $i-j \equiv h-k \pmod{2m+1}$. Since I is an I-matrix, there is only one 1 on each special diagonal. It follows that $-j \equiv -k \pmod{2m+1}$. Hence $j \equiv k \pmod{2m+1}$. Since we just proved that M is a permutation matrix, then $i \equiv h \pmod{2m+1}$. Hence $M_{i,j}$ and $M_{h,k}$ are the same entry of M. This completes the proof of the lemma. q.e.d.

Lemma VI.14: The transformation Δ has the properties that $\Delta = t \circ dc$, $[\Delta^2(I)]_{-n,k-n} = I_{k,n}$ and $\Delta^3 = 1$, where $\circ$ is the composition.

Proof: We have by (4), (3), and (1) of Definition VI.12 that

$$[\Delta(I)]_{n-k,-k} \overset{(4)}{=\!=} I_{k,n} \overset{(3)}{=\!=} [dc(I)]_{-k,n-k} \overset{(1)}{=\!=} [t(dc(I))]_{n-k,-k} =\!= [(t \circ dc)(I)]_{n-k,-k}.$$

Hence $\Delta = t \circ dc$.

By (4) of Definition VI.12, we have

$$I_{k,n} \overset{(4)}{=\!=} [\Delta(I)]_{n-k,-k} \overset{(4)}{=\!=} [\Delta(\Delta(I))]_{i,j} =\!= [\Delta^2(I)]_{i,j},$$

where $i = (-k)-(n-k) = -n$ and $j = -(n-k) = k-n$. Hence by (4) again,

$$I_{k,n} = [\Delta^2(I)]_{i,j} = [\Delta^2(I)]_{-n,k-n} \overset{(4)}{=\!=} [\Delta(\Delta^2(I))]_{k,n}$$
$$= [\Delta^3(I)]_{k,n},$$

i.e., $I = \Delta^3(I)$ and then $\Delta^3 = i$. q.e.d.

Lemma VI.15: For the transformation t, dr, and dc, we have $t^2 = (dr)^2 = (dc)^2 = i$.

Proof: By (1), (2), and (3) of Definition VI.12, we have

$$I_{k,n} \overset{(1)}{=\!=} [t(I)]_{n,k} \overset{(1)}{=\!=} [t(t(I))]_{k,n} =\!= [(t\circ t)(I)]_{k,n},$$

$$I_{k,n} \overset{(2)}{=\!=} [dr(I)]_{k-n,-n} \overset{(2)}{=\!=} [dr(dr(I))]_{k,n}$$
$$=\!= [(dr)^2(I)]_{k,n},$$

and

$$I_{k,n} \overset{(3)}{=\!=} [dc(I)]_{-k,n-k} \overset{(3)}{=\!=} [dc(dc(I))]_{k,n}$$
$$=\!= [(dc)^2(I)]_{k,n}.$$

Hence $t\circ t = dr\circ dr = dc\circ dc = i$. q.e.d.

Theorem VI.16: The set $\mathcal{G} = \{i, dr, dc, t, \Delta, \Delta^2\}$ of transformations acting on F_{2m} of all I-matrices of order 2m is a group under composition $\circ$.

Proof: Combining Lemmas VI.13, VI.14, and VI.15 and some other similar relations, we have the following table:

$\circ$	i	dr	dc	t	Δ	Δ^2
i	i	dr	dc	t	Δ	Δ^2
dr	dr	i	Δ^2	Δ	t	dc
dc	dc	Δ	i	Δ^2	dr	t
t	t	Δ^2	Δ	i	dc	dr
Δ	Δ	dc	t	dr	Δ^2	i
Δ^2	Δ^2	t	dr	dc	i	Δ

which satisfies all the conditions of a group. Hence $(\mathcal{G},\circ)$ is a group. q.e.d.

Since the group $(\mathcal{G},\circ)$ is of order 6, the only possible subgroups are of orders 2 or 3 and the singleton $\{i\}$. Let DR = $\{i,dr\}$, DC = $\{i,dc\}$, and S = $\{i,t\}$ denote the three subgroups of order 2, and TR = $\{i,\Delta,\Delta^2\}$ the subgroup of order 3.

Definition VI.17: An I-matrix I is said to have properties DRP, DCP, SP and TRP if and only if I is fixed by all transformations in the subgroups DR, DC, S, and TR, respectively. An I-matrix I is said to have the HP property if and only if I is fixed by (all transformations in) the whole group .

We then have the following lattice relationship between I-matrices with different properties. One sits above the other if it is fixed by more transformations.

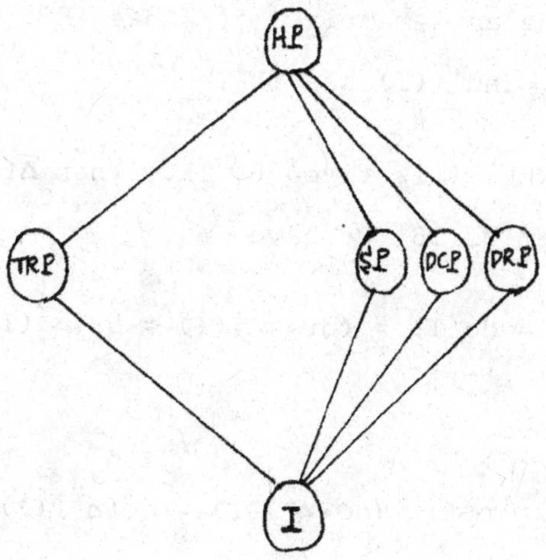

Figure 4

Examples:

We give six examples of I-matrices of order 6 to illustrate the above lattice. We list the permutations instead of the permutation I-matrices.

$I_6^{(1)}: \begin{pmatrix} 1 & 2 & 3 & 4 & 5 & 6 \\ 5 & 1 & 4 & 6 & 3 & 2 \end{pmatrix}$ $\quad I_6^{(4)}: \begin{pmatrix} 1 & 2 & 3 & 4 & 5 & 6 \\ 4 & 1 & 5 & 2 & 6 & 3 \end{pmatrix}$

$I_6^{(2)}: \begin{pmatrix} 1 & 2 & 3 & 4 & 5 & 6 \\ 3 & 6 & 2 & 5 & 1 & 4 \end{pmatrix}$ $\quad I_6^{(5)}: \begin{pmatrix} 1 & 2 & 3 & 4 & 5 & 6 \\ 2 & 4 & 6 & 1 & 3 & 5 \end{pmatrix}$

$I_6^{(3)}: \begin{pmatrix} 1 & 2 & 3 & 4 & 5 & 6 \\ 6 & 5 & 4 & 3 & 2 & 1 \end{pmatrix}$ $\quad I_6^{(6)}: \begin{pmatrix} 1 & 2 & 3 & 4 & 5 & 6 \\ 3 & 6 & 1 & 5 & 4 & 2 \end{pmatrix}$.

Here $I_6^{(6)}$ has HP, $I_6^{(2)}$ has TRP, $I_6^{(3)}$ has SP, $I_6^{(4)}$ has DCP and $I_6^{(5)}$ has DRP, $I_6^{(1)}$ is simply an I-matrix; it doesn't possess any of the properties defined in Definition VI.17.

Lemma VI.18: Let I be an I-matrix. If I has TRP, then each of dr(I), dc(I), and t(I) has TRP.

Proof: Since I has TRP, I is fixed by TR. Then $\Delta(I) = \Delta^2(I) = I$. By Theorem VI.16, we have

$$\Delta(dr(I)) = (\Delta \circ dr)(I) = dc(I) = (dr \circ \Delta^2)(I) = dr(\Delta^2(I)) = dr(I)$$

and

$$\Delta(dc(I)) = (\Delta \circ dc)(I) = t(I) = (dc \circ \Delta^2)(I) = dc(\Delta^2(I)) = dc(I)$$

and

$$\Delta(t(I)) = (\Delta \circ t)(I) = dr(I) = (t \circ \Delta^2)(I) = t(\Delta^2(I)) = t(I).$$

Similarly, we have $\Delta^2(dr(I)) = dr(I)$, $\Delta^2(dc(I)) = dc(I)$ and $\Delta^2(t(I)) = t(I)$. Thus dr(I), dc(I), and t(I) all have TRP. q.e.d.

Lemma VI.19: Let I be an I-matrix. If I has DRP, then dc(I) has SP and t(I) has DCP. If I has DCP, then dr(I) has SP and t(I) has DRP. If I has SP, then dr(I) has DCP and dc(I) has DRP.

Proof: We only show the first case when I has DRP, the remaining cases being similar.

Assume I has DRP. Then dr(I) = I. By Theorem VI.16,

$$t(dc(I)) = (t \circ dc)(I) = \Delta(I) = (dc \circ dr)(I) = dc(dr(I)) = dc(I).$$

Hence dc(I) has SP. On the other hand, we have

$$dc(t(I)) = (dc \circ t)(I) = \Delta^2(I) = (t \circ dr)(I) = t(dr(I)) = t(I).$$

Hence t(I) has DCP. This completes the proof. q.e.d.

By Lemmas VI.18 and VI.19, we conclude that TRP I-matrices exist on 2-cycles and SP, DCP, and DRP I-matrices exist on 3-cycles, i.e.:

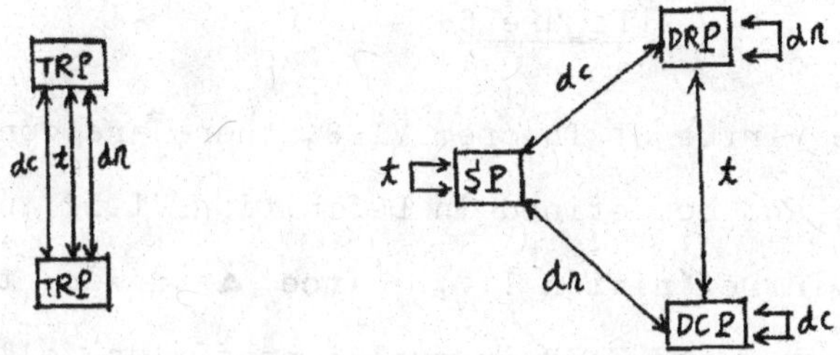

Figure 5

Moreover, since a HP I-matrix is fixed by all the transformations in $\mathfrak{G}$ and an I-matrix without any of the properties defined in Definition VI.17 is fixed only by i, we have the following:

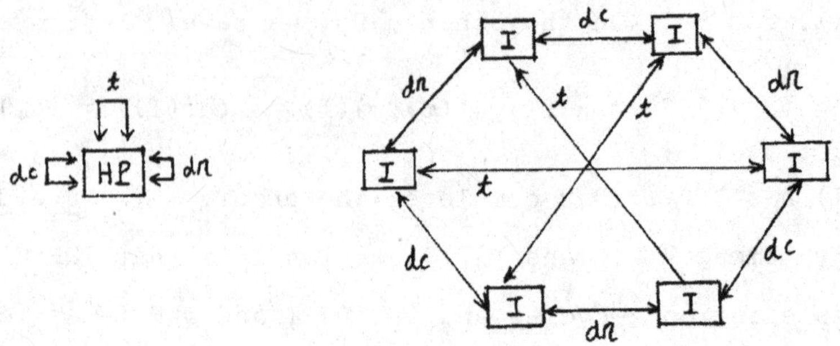

Figure 6

By the virtue of Theorem VI.8, there are connections between t, dr, and dc defined in Definition VI.12 and CD, RD, T defined in Definition I.7. Since $\Delta = dc \circ dr$, the relation between Δ and $T \circ RD$ in cyclic neofields follows. Let $A(N_v)$ be the I-matrix associated with the cyclic neofield N_v in the sense of Theorem VI.8. Hence $A(N_v)$ is of order v-2. We have the following theorem.

Theorem VI.20: Let N_v be a cyclic neofield of order v. Then we have the following:

$$\begin{cases} (1) & A(CD(N_v)) = t(A(N_v)) \\ (2) & A(RD(N_v)) = dr(A(N_v)) \\ (3) & A(T(N_v)) = dc(A(N_v)) \end{cases}$$

Proof: We only prove (1). The remaining two cases are similar. Let $1+a^k = a^n$ in N_v. Then in $N_v^* = CD(N_v)$, we have $1+a^n = a^k$. Hence in $A_{v-2}^* = A(N_v^*) = A(CD(N_v))$, the (n,k) entry is equal to 1.

On the other hand, let $1+a^k = a^n$ in N_v. Then in $A_{v-2} = A(N_v)$, we have that the (k,n) entry is equal to 1. Hence $t(A_{v-2}) = t(A(N_v))$, the (n,k) entry is equal to 1. It follows that matrices $A(CD(N_v))$ and $t(A(N_v))$ are identical to each other. This completes the proof. q.e.d.

According to the above theorem, We have the following diagram.

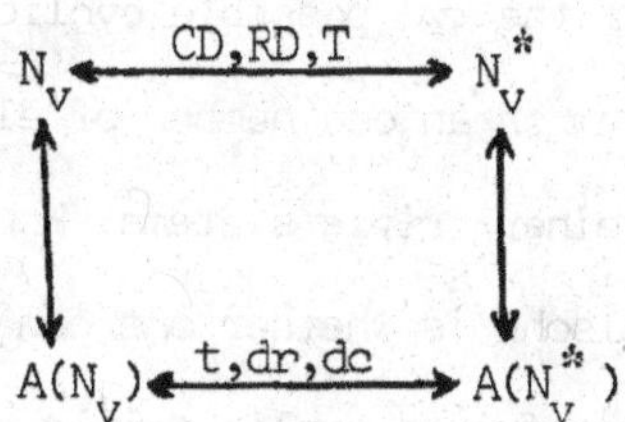

Figure 7

This gives us not only a broad combinatorial approach to cyclic neofield problems, but also a wide-range application of these non-associative algebras to various combinatorial problems.

Sec.4. Concluding Remarks

The current work has diversified applications (see for example Hsu [7] and [8]) which are not included because of the scope and the nature of this monograph. Here we illustrate only one of them.

A cyclic triple system (briefly CTS) is a pair (S,T) where S is a finite set and T is a non-empty collection of ordered 3-subsets of

S, called <u>cyclic triples</u>, such that every ordered pair of distinct elements of S is contained in exactly one cyclic triple of T (note that the cyclic triple a,b,c contains the ordered pair ab,bc and ca but not ba,cb or ac). The number $|S| = n$ is called the order of (S,T). It is proved by Mendelsohn [20] that the necessary and sufficient condition for the existence of a cyclic triple system of order n is $n \equiv 0$ or 1 (mod 3) with the exception of $n = 1$ and $n = 6$. In the case where $n \equiv 1$ or 3(mod 6), one can always construct a cyclic triple system by simply taking a Steiner triple system for the same value of n and using each triple twice, one in each of the two possible cyclic orders. But conversely, a cyclic triple system on an odd number of elements does not necessarily separate into two Steiner triple systems. However, a question of some interest raised by Mendelsohn is whether one can extract and in which ways a Steiner triple system from a cyclic triple system.

Mendelsohn ([19] and [20]) has shown that a cyclic triple system of order n constructed can be decomposed into two Steiner triple systems in at least $2^{(n-7)/6}$ ways if $n \equiv 1$ (mod 3), n is a prime; or $n = p^2$, where p is an odd prime which is congruent to 2 (mod 3). This left a list of unsettled orders: $n = 49,55,85,91,\cdots$ for which $n \equiv 1$ (mod 6). In a recent paper (Hsu [7]), the author constructed a family of cyclic triple system of order n by using SIP-neofields (i.e., type a) ii) XIP-neofields) of order n+1 for any positive n, and then showed that the cyclic triple systems of any order n, $n \equiv 1$ (mod 6) so constructed can always be decomposed into two Steiner triple systems in at least $2^{(n-7)/6}$ ways. Hence it completely solved the decomposition problems for the above listed orders.

We now state three questions arising from the current work.

Question 1: Johnsen and Storer [12] have shown that the number of non-isomorphic CIP-neofields of order $v = p^{\alpha}$ constructed by a special method goes to infinite with v. But in general, how many non-isomophic CIP-neofields can we get for a given order? Moreover, how many distinct (in the sense of presentation function) cyclic neofields can one get for a given order?

The number of cyclic neofields certainly increases with the order. A computer program run by MTS at The University of Michigan gives the number $\#(N_v)$ of cyclic neofields N_v up to order v = 12 (also see Appendix II).

v	4	5	6	7	8	9	10	11	12	...
$\#(N_v)$	1	2	3	8	19	64	225	928	3441	...

We have got some results on the enumeration of cyclic neofields for a given order, but to get an exact closed formula seems to be not easy. For example, if we have cyclic neofield N_v of even order v with $1+a = a^k$, then we have a different cyclic neofield with $1+a = a^{1-k}$. For in $T(N_v)$, we have $a+1 = a^k$. Then in $T^*(T(N_v))$ (see Definition II.1), we have $a^k(1+a) = a$, i.e., $1+a = a^{1-k}$. Hence in the enumeration of N_v for a given v, we only have to find the number of neofields in each case with $T(a) = a^2$, $T(a) = a^3$, $\cdots$, $T(a) = a^{v/2}$ when v is even, and with $T(1) = a$, $T(1) = a^2$, $\cdots$, $T(1) = a^{(v-1)/2}$ when v is odd.

Question 2: In Chapter III, type b) XIP-admissible partitions are constructed for Z_{v-1} with $v \equiv 2, 3, 5 \pmod 6$ and $v \equiv 4, 7 \pmod{18}$. Is there a general way to construct type b) XIP-admissible partitions for Z_{v-1} with the remaining orders except $v \equiv 0 \pmod 6$ or $v = 10$?

Here we display three type b) XIP-admissible partitions of Z_{v-1} with v = 16, 28 and 34.

<u>v = 16</u>,

$$S(1,7) = \{1, 7, 6, 14, 8, 9\}$$
$$S(7,3) = \{7, 3, 11, 8, 12, 4\}$$
$$S(3,1) = \{3, 1, 13, 12, 14, 2\}$$
$$S(9,11) = \{9, 11, 2, 6, 4, 13\}$$
$$S(5,10) = \{5, 10\}$$

<u>v = 28</u>,

$$S(1,5) = \{1, 5, 4, 26, 22, 23\}$$
$$S(3,11) = \{3, 11, 8, 24, 16, 19\}$$
$$S(5,17) = \{5, 17, 12, 22, 10, 15\}$$
$$S(24,4) = \{24, 4, 7, 3, 23, 20\}$$
$$S(14,2) = \{14, 2, 15, 13, 25, 12\}$$
$$S(20,6) = \{20, 6, 13, 7, 21, 14\}$$
$$S(2,8) = \{2, 8, 6, 25, 19, 21\}$$
$$S(26,10) = \{26, 10, 11, 1, 17, 16\}$$
$$S(9,18) = \{9, 18\}$$

<u>v = 34</u>,

$v-1 = 3 \times 11$. We take a number of order 10 modulo $v-1 = 33$, say 5. Then a type b) XIP-admissible partition π is given by

$$\Pi = \{S(5^0,5^2),\ S(5^2,5^4),\ S(5^4,5^6),\ S(5^6,5^8),\ S(5^8,1)\}$$
$$\cup \{S(5^1,5^3),\ S(5^3,5^5),\ S(5^5,5^7),\ S(5^7,5^9),\ S(5^9,5)\}.$$

Question 3: Theorem VI.10 asserts that if $p = 2n+1$ is an odd prime, then J^*_{2n} can be decomposed into the sum of $2n-1$ I-matrices A_{2n}. Is the theorem also true if $2n+1$ is a prime power or any other number? If the theorem is not true for non-prime power order, how big is the number of components we can get?

To answer the first part of Question 3, we check by exhaustion to see if J^*_8 can be decomposed into the sum of seven I-matrices A_8. A negative conclusion is reached. On the other hand, by Theorem VI.8, those $2n-1$ components arising is Theorem VI.10 correspond to $2n-1$ cyclic neofields N_{2n+2} of order $2n+2$. Now, we delete the first row and the first column of the addition table of these neofields and replace the main diagonal of each table by $1, a, a^2, \cdots, a^{2n}$. We then get $2n-1$ Latin squares Π_{2n+1} of order $2n+1$. It is not so difficult to check that the $2n-1$ Latin squares arising from Theorem VI.10 are mutually orthogonal. Hence we could get a complete set of mutually orthogonal Latin squares of prime order $p = 2n+1$.

The current work engenders several problems worthy of further study. We list three of them.

Question 4: Suppose that the multiplicative operation of a neofield is not cyclic. Can the theory of cyclic neofield be extended in a slightly different way to abelian neofields, or even to non-abelian neofields?

Had we started Chapter I with slightly more general definitions, we should have arrived, via the same type of development, at a theory for abelian neofields which closely parallels the previous theory for cyclic neofields. Hence the first part of Question 4 must be answered affirmatively. Along this line but only in the case of inverse property neofield, Tannenbaum [24] showed that an abelian inverse property neofield (AIP-neofield) of even order v exists if and only if $v \equiv 2$ or 4 (mod 6) and the multiplicative group of non-zero elements is not the cyclic group of order 9. He also showed that the existence of an odd order v AIP-neofield with multiplicative group of non-zero elements (let $v = w+1$) has the following necessary conditions (a) wheneven A is written in the form $A = C_{w_1} \times C_{w_2} \times \cdots \times C_{w_t}$, where w_i divides w_{i-1}, then w_1 must be the only even term among the w_i's; (b) $w \not\equiv 14, 20$ (mod 24). Consequently, AIP-neofields are used to study "abelian" Steiner triple systems.

Question 5: Since the finite fields are just CIP-neofields with associative addition, can the method of construction of general cyclic neofields given here be specialized to provide construction for finite fields?

It is worth noting that the constructions given here develop the additive structure in a cyclic neofield in terms of a multiplicative structure, while the Galois construction of finite fields determines the multiplicative structure in terms of the additive.

Question 6: Can symmetric block design be found in non-prime-power CIP-neofields or other general cyclic neofields?

It is established (Doner[4] and Johnsen and Storer[13])

that Lehmer difference sets in the additive loop of some CIP-neofields of prime-power order can give rise to symmetric block designs. It is shown that (Doner[4]) the fourth powers in a CIP-neofield of order v can form a loop difference set only if v = 5+96t (t = 1,2, ⋯) or v = 37+96t (t = 0, 1, ⋯). So far, no symmetric block design has been found yet in non-prime-power CIP-neofields or other general cyclic neofields. On the other hand, the cyclotomy for cyclic neofields can be developed along lines completely independent of the classical study of field cyclotomy (see Storer[23]).

Appendix I

Type a) XIP-Admissible Partitions

In this Appendix, a type a) XIP-admissible partition of Z_{v-1}^{*} is given. By Theorem I.41, if v is even, $v \equiv 2$ or 4 (mod 6). By Theorem II.15, if v is odd, $v \not\equiv 15,21 \pmod{24}$.

First of all, when v is even, we divide into 8 cases. We consider the triples $\{k, n, n-k\}$ instead of the sextuples $\{k, n, n-k, -k, -n, k-n\}$ because the last three numbers can be obtained accordingly.

Case 1: $v = 24m+2$, $m \geq 1$.

We must partition Z_{24m+1}^{*} into 4m sextuples $\{k, n, n-k, -k, -n, k-n\}$. We provide a general construction for $m \geq 1$, yielding 4m triples, with n-k ranging over $1, \cdots, 4m$. These 4m triples will be extended to sextuples which exactly partition Z_{24m+1}^{*}.

First, pair n and k by the scheme

k	n	n-k
4m+1	8m	4m-1
4m+2	8m-1	4m-3
.	.	.

(continued)

$\underline{k}$	$\underline{n}$	$\underline{n-k}$
.	.	.
.	.	.
6m-1	6m+2	3 ,

yielding by a "nesting" procedure a triple $\{k, n, n-k\}$ for each odd n-k from 3 to 4m-1.

Next, select the triples

$\underline{k}$	$\underline{n}$	$\underline{n-k}$
6m	10m	4m
6m+1	8m+1	2m
11m	11m+1	1 ,

and finally, another nesting given by

$\underline{k}$	$\underline{n}$	$\underline{n-k}$
8m+2	12m	4m-2
8m+3	12m-1	4m-4
.	.	.
.	.	.
.	.	.
9m	11m+2	2m+2
9m+1	11m-1	2m-2
9m+2	11m-2	2m-4
.	.	.
.	.	.
.	.	.
10m-1	10m+1	2 .

The triples so enumerated have n-k ranging over 1, 2, $\cdots$, 4m, while the corresponding pairs k,n range over

$\{4m+1, \cdots, 12m\}$. Extended to sextuples, then, we have a type a) XIP-admissible partition of Z^*_{24m+1} .

Caes 2: $v = 24m+8$, $m \geq 2$.

This construction is similar to the last: We partition $\{1, 2, \cdots, 12m+3\}$ into triples $\{k, n, n-k\}$ with n-k ranging over 1, 2, ⋯, 4m+1.

First, the nesting procedure

k	n	n-k
8m+3	12m+3	4m
8m+4	12m+2	4m-2
.	.	.
.	.	.
.	.	.
10m+2	10m+4	2 ,

then three triple

k	n	n-k
6m+2	10m+3	4m+1
6m+3	8m+2	2m-1
5m+2	5m+3	1 ,

and finally the nesting procedure

k	n	$n-k$
4m+2	8m+1	4m-1
4m+3	8m	4m-3
.	.	.
.	.	.
.	.	.
5m+1	7m+2	2m+1
5m+4	7m+1	2m-3
5m+5	7m	2m-5
.	.	.
.	.	.
.	.	.
6m+1	6m+4	3 .

The number n-k ranges over 1, 2, $\cdots$, 4m+1, and that the pairs k, n exhaust $\{4m+2, \cdots, 12m+3\}$.

This construction requires that $m \geq 2$. For $m = 0$, the sextuple $\{1, 5, 4, 6, 2, 3\}$ would be an example, and for m = 1, we take the following type a) XIP-admissible partition of Z_{31}^{*} .

$$(1,18) = \{1, 18, 17, 30, 13, 14\}$$
$$(2,28) = \{2, 28, 26, 29, 3, 5\}$$
$$(4,10) = \{4, 10, 6, 27, 21, 25\}$$
$$(7,22) = \{7, 22, 15, 24, 9, 16\}$$
$$(8,19) = \{8, 19, 11, 23, 12, 20\}.$$

Case 3: $v = 24m+14, \quad m \geq 2.$

A slight modification of procedure becomes necessary in this case. We partition $\{1, 2, \cdots, 12m+5\} \cup \{12m+7\}$ into triples $\{k, n, n-k\}$ with n-k ranging over 1, 2,$\cdots$, 4m+2, and the triples extended to sextuples obviously provide an admissible partition of Z^*_{24m+13}.

We begin with the nesting

k	n	n-k
4m+3	8m+5	4m+2
4m+4	8m+4	4m
.	.	.
.	.	.
.	.	.
6m+3	6m+5	2 ,

following by three triples

k	n	n-k
6m+4	10m+5	4m+1
10m+4	12m+7	2m+3
11m+6	11m+7	1 ,

and finally a nesting procedure

k	n	n-k	
8m+6	12m+5	4m-1	(continued)

k	n	n-k
8m+7	12m+4	4m-3
.	.	.
.	.	.
.	.	.
9m+3	11m+8	2m+5
9m+4	11m+5	2m+1
9m+5	11m+4	2m-1
.	.	.
.	.	.
.	.	.
10m+3	10m+6	3 .

The triples so determined have n-k ranging over 1, 2, ···, 4m+2 and the pairs k, n exhausting $\{4m+3, 4m+4, \cdots, 12m+5\} \cup \{12m+7\}$.

For v = 14, we have $\{\{3, 4, 1, 12, 11, 10\}, \{5, 7, 2, 8, 6, 11\}\}$. For v = 38, we have six sextuples:

$$\{10, 11, 1, 27, 26, 36\}, \{8, 12, 4, 29, 25, 33\}$$
$$\{15, 17, 2, 22, 20, 35\}, \{9, 14, 5, 28, 23, 32\}$$
$$\{16, 19, 3, 21, 18, 34\}, \{7, 13, 6, 30, 24, 31\}.$$

Case 4: v = 24m+20, m ≥ 1.

This case is similar to case 3. We begin with a

nesting procedure

k	n	n-k
4m+4	8m+6	4m+2
4m+5	8m+5	4m
.	.	.
.	.	.
.	.	.
6m+4	6m+6	2 ,

followed with the three triples

k	n	n-k
6m+5	10m+8	4m+3
10m+9	12m+10	2m+1
9m+7	9m+8	1 ,

and finally a nesting procedure

k	n	n-k
8m+7	12m+8	4m+1
8m+8	12m+7	4m-1
.	.	.
.	.	.
.	.	.
9m+6	11m+9	2m+3

(continued)

$\underline{k}$	$\underline{n}$	$\underline{n-k}$
9m+9	11m+8	2m-1
9m+10	11m+7	2m-3
.	.	.
.	.	.
.	.	.
10m+7	10m+10	3 .

This partitions $\{k, n, n-k\}$ into triples with n-k ranging over 1, 2, ... , 4m+3, and the pairs k, n exhausting $\{4m+4, \cdots, 12m+8\} \cup \{12m+10\}$.

Not included in the above construction is the case v = 20, for which triples $\{k, n, n-k\}$ with n-k ranging over 1, 2, 3 is given by

$$\{5, 6, 1\} , \{8, 10, 2\} , \{4, 7, 3\}.$$

<u>Case 5</u>: v = 24m+4, m $\geq$ 1.

For this case, we must partition Z_{24m+3}- $\{0, (v-1)/3, (2(v-1))/3\}$ into sextuples. In particular, we wish to partition

$$\{1, 2, \cdots, 12m+1\} - \{8m+1\}$$

into triples $\{k, n, n-k\}$ with n-k ranging over 1, 2, $\cdots$, 4m.

We begin with a nesting

$\underline{k}$	$\underline{n}$	$\underline{n-k}$
8m+3	12m+1	4m-2
8m+4	12m	4m-4
.	.	.
.	.	.
.	.	.
10m+1	10m+3	2 ,

followed by three special triples

$\underline{k}$	$\underline{n}$	$\underline{n-k}$
6m+2	10m+2	4m
6m+1	8m+2	2m+1
5m	5m+1	1 ,

and a nesting

$\underline{k}$	$\underline{n}$	$\underline{n-k}$	
4m+1	8m	4m-1	
4m+2	8m-1	4m-3	
.	.	.	
.	.	.	
.	.	.	
5m-1	7m+2	2m+3	
5m+2	7m+1	2m-1	(continued)

$\underline{k}$	$\underline{n}$	$\underline{n-k}$
5m+3	7m	2m-3
.	.	.
.	.	.
.	.	.
6m	6m+3	3 .

The resulting triples have n-k ranging over 1, 2, $\cdots$, 4m, and the pairs k, n exhausting

$$\{4m+1, \cdots, 12m+1\} - \{8m+1\} .$$

Not included in the above construction is the case v = 4, for which we don't have partition into sextuples. Hence GF(4) is the only CIP-neofield and SIP-neofield (see Theorem II.18).

Case 6: v = 24m+22, $m \geq 1$.

We partition $\{1, 2, \cdots, 12m+10\} - \{8m+7\}$ into triples $\{k, n, n-k\}$ with n-k ranging over 1, 2,$\cdots$, 4m+3.

First, we use the nesting

$\underline{k}$	$\underline{n}$	$\underline{n-k}$
8m+8	12m+10	4m+2
8m+9	12m+9	4m
.	.	.
.	.	.

(continued)

k	n	n-k
.	.	.
10m+8	10m+10	2 ,

and then three triples

k	n	n-k
6m+6	10m+9	4m+3
6m+5	8m+6	2m+1
5m+4	5m+5	1 ,

followed by a nesting

k	n	n-k
4m+4	8m+5	4m+1
4m+5	8m+4	4m-1
.	.	.
.	.	.
.	.	.
5m+3	7m+6	2m+3
5m+6	7m+5	2m-1
.	.	.
.	.	.
.	.	.
6m+4	6m+7	3 .

This provide a partition of $\{1, 2, \ldots, 12m+10\} - \{8m+7\}$

into triples $\{k, n, n-k\}$ with n-k ranging over 1, 2, ⋯, 4m+3, and the pairs k, n exhausting

$$\{4m+4, \cdots, 12m+10\} - \{8m+7\} ,$$

Not covered by the above construction is the case v = 22, for which we supply the example with n-k ranging over 1, 2, 3, and the triples are given by

$$\{4, 5, 1\} , \{8, 10, 2\} , \{6, 9, 3\} .$$

Case 7: $v = 24m+10, \quad m \geq 2$.

In this case, we partition $\{1, 2, \cdots, 12m+3, 12m+5\} - \{8m+3\}$ into triples $\{k, n, n-k\}$ with n-k ranging over 1, 2, ⋯, 4m+1. We begin with a nesting

k	n	n-k
4m+2	8m+2	4m
4m+3	8m+1	4m-2
.	.	.
.	.	.
.	.	.
6m+1	6m+3	2 ,

followed by special triples

k	n	$n-k$
6m+2	10m+3	4m+1
10m+2	12m+5	2m+3
11m+4	11m+5	1 ,

and finally the nesting

k	n	$n-k$
8m+4	12m+3	4m-1
8m+5	12m+2	4m-3
.	.	.
.	.	.
.	.	.
9m+1	11m+6	2m+5
9m+2	11m+3	2m+1
.	.	.
.	.	.
.	.	.
10m+1	10m+4	3 .

The triples so determined have n-k ranging over 1, 2, $\cdots$, 4m+1, with the pairs k, n exhausting

$$\{4m+2, \cdots, 12m+3, 12m+5\} - \{8m+3\} .$$

The construction fails for m = 1, so we supply an example obtained by trial, with n-k ranging over 1, 2, $\cdots$, 5 and the triples given by

$$\{6, 7, 1\}, \{8, 10, 2\}, \{12, 15, 3\},$$
$$\{13, 17, 4\}, \{9, 14, 5\}.$$

Note that when $m = 0$, there is no type a) XIP-admissible partition of Z_9^*.

Case 8: $v = 24m+16$, $m \geq 1$.

In this case, we partition

$$(\{1, \cdots, 12m+7\} \cup \{13m+8\}) - \{8m+5, 11m+7\}$$

into triples $\{k, n, n-k\}$ with $n-k$ ranging over $1, 2, \cdots, 4m+2$. Since $13m+8 \equiv -(11m+7) \pmod{(24m+15)}$, when these triples are extended to sextuples, we get an admissible partition of Z_{24m+15}^*.

We begin with the nesting procedure

k	n	n-k
4m+3	8m+3	4m
4m+4	8m+2	4m-2
.	.	.
.	.	.
.	.	.
6m+2	6m+4	2 ,

followed by two triples

k	n	n-k
6m+3	8m+4	2m+1
9m+6	13m+8	4m+2 ,

and a nesting

k	n	n-k
8m+6	12m+7	4m+1
8m+7	12m+6	4m-1
.	.	.
.	.	.
.	.	.
9m+5	11m+8	2m+3
9m+7	11m+6	2m-1
9m+8	11m+5	2m-3
.	.	.
.	.	.
.	.	.
10m+6	10m+7	1 .

The construction does not provide for $m = 0$. In that case we have triples $\{1, 4, 3\}$ and $\{2, 8, 6\}$.

Secondly, when v is odd, we divide into 3 cases. We supply a partition of $Z_{v-1}^{*} = Z_{v-1} - \{(v-1)/2\}$ of the correct form and treat cases according to congrunce classes of v modulo 3.

Case 9: $v \equiv 2 \pmod 3$.

Let $v = 3s+2$, $s \geq 1$ odd, and take $m = v-1 = 3s+1$; then we must obtain a partition of $Z_m - \{m/2\}$ consisting of one triple $\{0, r, -r\}$ for $r \neq 0$, and $(s-1)/2$ sextuples of the form

$$\{k, n, n-k, -k, -n, k-n\} .$$

Since the negative of any element in the set of residues $\{1, 2, \cdots, (m-2)/2\}$ (mod m) is in $\{(m+2)/2, \cdots, m-1\}$, it suffices, in fact, to find a partition of $\{1, \cdots, (m-2)/2\}$ into $(s-1)/2$ triples of the form $\{k, n, n-k\}$, and a single element r not included in those triples. In fact, we find such triples in such a way that n-k runs over the elements in $\{1, \cdots, (s-1)/2\}$

First, from the residues $\{(s+1)/2, \cdots, (s-1)\}$ we select pairs k, n as follows:

k	n	n-k
$(s+1)/2$	$(s-1)$	$(s-3)/2$
$(s+3)/2$	$(s-2)$	$(s-7)/2$
.	.	.
.	.	.
.	.	. .

We proceed in this fashion, selecting pairs k, n as the extreme-most elements of $\{(s+1)/2, \cdots, (s-1)\}$ not yet used, until all but one, or all elements of $\{(s+1)/2, \ldots, (s-1)\}$ have been used. These pairs k, n provided differences,

$$(s-3)/2,\ (s-7)/2,\ \cdots,\ \begin{cases} 2, & s \equiv 3 \pmod 4 \\ 1, & s \equiv 1 \pmod 4), \end{cases}$$

i.e., the differences n-k run over all odd or all even residues in $\{1, 2, \cdots, (s-1)/2\}$.

Next, from the set of residues

$$\{s, \cdots, (m-2)/2 = (3s-1)/3\} ,$$

select pairs k, n in the same fashion, getting

k	n	n-k
s	(3s-1)/2	(s-1)/2
s+1	(3s-3)/2	(s-5)/2
.	.	.
.	.	.
.	.	.

proceeding inward from the extremes of $\{s, \ldots, (3s-1)/2\}$ until all but one, or all elements of that set are exhausted. The set of differences n-k in this case is

$$(s-1)/2,\ (s-5)/2,\ \cdots,\ \begin{cases} 1, & s \equiv 3 \pmod 4) \\ 2, & s \equiv 1 \pmod 4) \end{cases}$$

i.e., we have selected $(s-1)/2$ pairs k, n from the set $\{(s+1)/2, \cdots, (s-1)\} \cup \{s, \cdots, (m-2)/2\}$ such that differences n-k constitute the set $\{1, \cdots, (s-1)/2\}$. Hence the $(s-1)/2$ triples k, n, n-k exhaust all of $\{1, 2, \cdots, (m-2)/2\}$ except for one element, which we designate as r; thus we get a partition of $Z_m - \{m/2\}$ consisting of $(s-1)/2$ sextuples

$$\{k, n, n-k, -k, -n, k-n\}$$

and the triple $\{0, r, -r\}$; this is a type a) XIP-admissible partition of Z_{v-1}.

<u>Case 10</u>: $v \equiv 1 \pmod 3$.

Let $v = 3s+1$, $s \geq 2$ even, and take $m = v-1 = 3s$; then we must obtain a partition of $Z_m - \{m/2\}$ consisting of one triple

$$\{0, r, -r\},$$

the pair

$$\{m/3 = s, 2m/3 = 2s\},$$

and $(s/2)-1$ sextuples

$$\{k, n, n-k, -k, -n, k-n\}.$$

As above, we obtain triples $\{k, n, n-k\}$ from

$$\{1, \cdots, (3s-2)/2\} - \{s\} ,$$

with the differences n-k ranging over the set $\{1, 2, \cdots, (s/2)-1\}$. The interval $\{s/2, \cdots, s-1\}$ provides pairs k, n such that we have

k	n	n-k
(s/2)	s-1	(s/2)-1
(s/2)+1	s-2	(s/2)-3
.	.	.
.	.	.
.	.	.

yielding the differences n-k given by

$$(s/2)-1, (s/2)-3, \cdots, \begin{cases} 1, & s \equiv 0 \pmod 4 \\ 2, & s \equiv 2 \pmod 4 \end{cases} ,$$

and the interval $\{s+1, \cdots, (3s-2)/2\}$ provides pairs k, n such that we have

k	n	n-k
s+1	(3s-2)/2	(s/2)-2
s+2	(3s-4)/2	(s/2)-4
.	.	.
.	.	.
.	.	.

yielding the differences n-k given by

$$(s/2)-2,\ (s/2)-4,\ \cdots,\ \begin{cases} 1, & s \equiv 2 \pmod 4 \\ 2, & s \equiv 0 \pmod 4 \end{cases} .$$

The triples chosen above exhaust $\{1, 2, \cdots, (3s-2)/2\}$ except for s and one element r, which are used to form the pair $\{s, 2s = -s\}$, and the triple $\{0, r, -r\}$. The sextuples and triple correspond to additive structure as shown under the case $v \equiv 2 \pmod 3$, and the pair corresponds to the additive structure $\{(s, s/2), (2s, (5s)/2)\}$.

<u>Case 11</u>: $v \equiv 0 \pmod 3$

By Theorem II.15, we know that $v \equiv 3$ or $9 \pmod{24}$.

Consider first $v = 24s+3$, $s \geq 0$, and take $m = v-1$. We must partition Z_m^* into a singletion $\{0\}$, and 4s sextuples. As before, we choose triples $\{k, n, n-k\}$ from $\{1, \cdots, (m/2)-1 = 12s\}$ such that the differences n-k range over $\{1, \cdots, 4s\}$.

We obtain the odd differences 4s-1, 4s-3, ⋯, 1 by selecting pairs k, n in $\{4s+1, \cdots, 8s\}$ by the scheme

k	n	n-k
4s+1	8s	4s-1
4s+2	8s-1	4s-3
.	.	.
.	.	.
.	.	.

Clearly the pairs k, n exactly exhaust $\{4s+1, \cdots, 8s\}$. The technique used so far quite obviously fails if applied to $\{8s+1, \cdots, 12s\}$ to obtain pairs k, n such that the differences n-k run over 2, 3, $\cdots$, 4s, so we proceed instead by first selecting pairs k, n according to the following:

k	n	n-k
10s	10s+2	2
10s-1	10s+3	4
.	.	.
.	.	.
.	.	.
10s-(s-2)	10s+s	2s-2
10s-s	10s+(s+2)	2s+2
10s-(s+1)	10s+(s+3)	2s+4
.	.	.
.	.	.
.	.	.
10s-(2s-2)	10s+2s	4s-2 .

Four elements of $\{8s+1, \cdots, 12s\}$ are not used in the above association, viz.,

$$8s+1,\ 10s-(s-1) = 9s-1,\ 10s+1,\ 10s+(s+1) = 11s+1,$$

and two differences n-k do not occur above, viz.,

$n-k = 2s$, or $n-k = 4s$.

We judiciously combine these unused elements, but with the liberty of using an element not in $\{8s+1, \cdots, 12s\}$, namely, $14s-(s-1) = 13s+1$, in place of $11s+1$. Since, in fact, $11s+1 = -(13s+1)$, the elements of $\{1, 2, \cdots, 12s\}$ are still exactly exhausted when the triples k, n, n-k obtained are extended to sextuples. In particular, then, we obtain the differneces 2s, 4s via the correspondence

k	n	n-k
8s+1	10s+1	2s
9s+1	13s+1	4s.

The resulting triples exhaust ($\{1, 2, \cdots, 12s\} \cup \{13s+1\}$) $- \{-(13s+1)\}$, so when extended to sextuples, provide a partition of $Z_m - \{0, m/2\}$.

For $v = 24s+9$, $s \geq 0$, and $m = v-1$, we must partition $Z_m - \{m/2\}$ into a singleton $\{0\}$, and 4s+1 sextuples. We proceed again to find triples $\{k, n, n-k\}$ such that the differences n-k run over the set $\{1, \cdots, 4s+1\}$, and the pairs k, n lie (with one exception) in $\{4s+2, \cdots, 12s+3\}$.

First, pairs k, n are selected from $\{4s+2, \cdots, 8s+3\}$ to obtain the odd differences $4s+1, 4s-1, \cdots, 1$ in $1, 2, \cdots, 4s+1$ by the correspondence

$\underline{k}$	$\underline{n}$	$\underline{n-k}$
$4s+2$	$8s+3$	$4s+1$
$4s+3$	$8s+2$	$4s-1$
.	.	.
.	.	.
.	.	. .

Clearly, the pairs k, n so chosen exhaust $\{4s+2, \cdots, 8s+3\}$.

The procedure for obtaining the differences 2, $\cdots$, 4s from pairs in $\{8s+4, \cdots, 12s+3\}$ is similar to the procedure for $v \equiv 3 \pmod{24}$, the particular correspondence being

$\underline{k}$	$\underline{n}$	$\underline{n-k}$
$10s+3$	$10s+5$	2
.	.	.
.	.	.
.	.	.
$(10s+3)-(s-2)$	$(10s+5)+(s-2)$	$2s-2$
$(10s+3)-s$	$(10s+5)+s$	$2s+2$
.	.	.
.	.	.
.	.	.
$(10s+3)-(2s-2)$	$(10s+5)+(2s-2)$	$4s-2$.

The elements of $\{8s+4, \cdots, 12s+3\}$ not used in the above solution are $8s+4$, $(10s+3)-(s-1) = 9s+4$, $10s+4$,

$(10s+5)+(s-1) = 11s+4$, and we must still obtain the differences $2s$, $4s$. Again, we use $13s+4 = -(11s+4)$ in this selection of k, n, as follows:

k	n	n-k
$8s+4$	$10s+4$	$2s$
$9s+4$	$13s+4$	$4s$.

Hence the triples $\{k, n, n-k\}$ chosen from above exhaust $(\{1, 2, \cdots, 12s+3\} \cup \{13s+4\}) - \{11s+4\}$, so when extended to sextuples, all of $Z_m - \{m/2\}$ is exhausted.

Appendix II

Cyclic Neofield N_v, $v < 11$

In this Appendix, we list cyclic neofields N_v of all orders $5 \leq v \leq 10$. Note that this Fortran program is valid for any finite v. (see the table in Question 1 of Chapter VI.4). Also, the program can be modified to print only the number of cyclic neofields for any given order in an efficient way.

Each cyclic neofield is identified by the exponents of its presentation function. The table is sorted in the order of increasing values of v and the number N of cyclic neofield N_v (with distinct presentation functions) appears at the end of the printing of all cyclic neofields of order v. Each cyclic neofield is indicated as a row vector. For example, a print-out: 3 6 1 5 4 2 under the even order v = 8 would indicate the cyclic neofield N_8 with presentation function:

x	0	1	a	a^2	a^3	a^4	a^5	a^6
$1+x$	1	0	a^3	a^6	a	a^5	a^4	a^2

,

while a print-out: 1 3 5 2 4 under the odd order v = 7 would indicate the cyclic neofield N_7 with presentation function:

x	0	1	a	a^2	a^3	a^4	a^5
$1+x$	1	a	a^3	a^5	0	a^2	a^4

MICHIGAN TERMINAL SYSTEM FORTRAN G(21.8) MAIN 04-09

```
      IMPLICIT INTEGER (A-Z)  (
      DIMENSION A(50),B(50),C(50),E(50)
      DO 730 V=5,16
      WRITE (6,800)V
      VM1=V-1
      DO 80 I=1,VM1
      C7=0
      E(I)=1
      C(I)=1
      A(I)=I
   80 CONTINUE
      VD2=V/2*2
      IF(VD2.EQ.V) GO TO 100
      NV1=(V-1)/2
      DO 90 I=1,NV1
      A(I)=I-1
   90 CONTINUE
  100 P=1
  110 Q=1
  120 IF(C(Q).EQ.0) GO TO 190
      Y=Q-A(P)
  710 IF(Y.GT.0) GO TO 720
      Y=Y+V-1
      GO TO 710
  720 IF(E(Y).EQ.0) GO TO 190
      B(P)=Q
      C(Q)=0
      E(Y)=0
      P=P+1
      VM3=V-1
      IF (P.EQ.VM3) GO TO 510
      GO TO 110
  190 Q=Q+1
      VM2=V-2
      IF (Q.LE.VM2) GO TO 120
      P=P-1
      IF (P.GT.0) GO TO 230
      GO TO 730
  230 Q=B(P)
      C(Q)=1
      Y=Q-A(P)
  410 IF(Y.GT.0) GO TO 420
      Y=Y+V-1
      GO TO 410
  420 E(Y)=1
      E(Y)=1
      GO TO 190
  510 WRITE (6,600) (B(I),I=1,VM2)
      C7=C7+1
  530 P=P-1
      GO TO 230
  730 WRITE(6,900) C7
      CONTINUE
  500 FORMAT (1X)
  600 FORMAT (1X,20I4)
```

```
              900 FORMAT (1X,'N=',I7////)
              800 FORMAT (1X,'V=',I2)
                  RETURN
                  END
  *OPTIONS IN EFFECT*  ID,EBCDIC,SOURCE,NOLIST,NODECK,LOAD,NOMAP
  *OPTIONS IN EFFECT*  NAME = MAIN    , LINECNT =        57
  *STATISTICS*     SOURCE STATEMENTS =         59,PROGRAM SIZE =     2042
  *STATISTICS*  NO DIAGNOSTICS GENERATED
NO ERRORS IN MAIN

NO STATEMENTS FLAGGED IN THE ABOVE COMPILATIONS.
EXECUTION TERMINATED    22:44:41   T=.256  RC=0        $.18
```

```
$RUN -LOAD T=1
EXECUTION BEGINS    22:44:42
V= 5
   1    3    2
   3    2    1
N=       2



V= 6
   2    4    1    3
   3    1    4    2
   4    3    2    1
N=       3



V= 7
   1    3    5    2    4
   1    5    4    3    2
   2    4    1    5    3
   2    5    3    1    4
   4    2    5    3    1
   4    3    1    5    2
   5    2    4    1    3
   5    4    3    2    1
N=       8



V= 8
   2    4    6    1    3    5
   2    5    1    6    4    3
   2    6    5    3    1    4
   3    1    6    5    2    4
   3    5    2    1    6    4
   3    6    1    5    4    2
   3    6    2    5    1    4
   3    6    4    2    1    5
   4    1    5    2    6    3
   4    3    1    6    2    5
   4    6    2    5    3    1
   5    1    4    6    3    2
   5    3    1    6    4    2
   5    3    2    6    1    4
   5    3    6    2    4    1
   5    4    1    3    6    2
   6    3    5    1    4    2
   6    4    2    5    1    3
   6    5    4    3    2    1
N=      19
```

```
V=  9
   1    3    5    7    2    4    6
   1    3    7    6    4    2    5
   1    4    6    2    7    3    5
   1    4    7    5    3    2    6
   1    6    4    7    3    5    2
   1    6    5    2    7    4    3
   1    7    4    6    2    5    3
   1    7    6    5    4    3    2
   2    4    1    7    6    3    5
   2    4    7    1    6    5    3
   2    5    1    6    3    7    4
   2    5    7    4    3    1    6
   2    6    3    7    4    1    5
   2    6    5    1    4    7    3
   2    7    3    6    1    5    4
   2    7    5    4    1    3    6
   3    2    6    1    7    5    4
   3    2    7    5    1    4    6
   3    5    4    1    2    7    6
   3    5    7    2    6    4    1
   3    6    1    4    7    2    5
   3    6    4    2    1    7    5
   3    7    1    5    6    2    4
   3    7    6    4    2    5    1
   4    2    1    6    7    3    5
   4    2    5    1    7    3    6
   4    2    7    5    3    1    6
   4    2    7    6    3    5    1
   4    3    1    6    2    7    5
   4    3    5    1    2    7    6
   4    3    7    1    6    5    2
   4    3    7    2    6    1    5
   4    6    1    5    3    7    2
   4    6    3    1    7    5    2
   4    6    3    2    7    1    5
   4    6    5    2    3    7    1
   4    7    1    5    6    3    2
   4    7    3    5    2    1    6
   4    7    3    6    2    5    1
   4    7    5    2    6    3    1
   5    2    1    6    7    4    3
   5    2    4    7    3    1    6
   5    3    1    7    6    4    2
   5    3    6    1    4    7    2
   5    4    3    1    7    2    6
   5    4    6    2    3    7    1
   5    7    3    6    4    2    1
   5    7    4    2    6    1    3
```

6	2	5	7	4	3	1
6	2	7	5	4	1	3
6	3	5	2	1	7	4
6	3	7	4	1	5	2
6	4	1	5	2	7	3
6	4	3	7	2	5	1
6	5	1	4	7	3	2
6	5	3	2	7	1	4
7	2	4	6	1	3	5
7	2	6	5	3	1	4
7	3	5	1	6	2	4
7	3	6	4	2	1	5
7	5	3	6	2	4	1
7	5	4	1	6	3	2
7	6	3	5	1	4	2
7	6	5	4	3	2	1

N= 64

V=10

2	1	6	8	7	3	5	4
2	1	6	8	7	4	3	5
2	1	7	6	8	3	5	4
2	1	7	6	8	4	3	5
2	4	6	8	1	3	5	7
2	4	7	1	8	5	3	6
2	4	8	7	3	1	6	5
2	5	1	8	7	3	6	4
2	5	7	1	3	8	6	4
2	5	7	1	4	8	3	6
2	5	7	3	1	8	4	6
2	5	8	1	7	4	6	3
2	5	8	3	7	1	4	6
2	5	8	6	3	1	4	7
2	6	1	7	4	8	3	5
2	6	5	1	8	4	3	7
2	6	8	3	7	4	1	5
2	6	8	7	3	5	4	1
2	6	8	7	4	3	5	1
2	7	1	6	8	5	4	3
2	7	5	1	8	4	6	3
2	7	5	3	8	1	4	6
2	7	5	8	4	3	1	6
2	7	6	1	4	8	5	3
2	7	6	8	3	5	4	1
2	7	6	8	4	3	5	1
2	8	5	7	1	4	6	3
2	8	5	7	3	1	6	4
2	8	5	7	4	1	3	6
2	8	6	3	7	1	5	4
2	8	7	6	3	5	1	4
3	1	6	8	2	7	5	4
3	1	7	5	8	2	4	6
3	1	8	7	6	4	2	5
3	5	1	8	2	7	6	4
3	5	1	8	6	2	4	7
3	5	2	8	1	7	4	6
3	5	7	1	4	2	8	6
3	5	8	1	4	7	2	6
3	5	8	1	6	4	2	7
3	5	8	2	6	1	4	7
3	6	1	5	8	2	4	7
3	6	1	7	2	5	8	4
3	6	1	7	4	2	8	5
3	6	2	1	8	7	5	4
3	6	2	7	1	4	8	5

3	6	4	1	8	2	5
3	6	8	1	4	7	5
3	6	8	2	4	7	1
3	6	8	5	2	4	1
3	7	1	8	6	5	4
3	7	2	5	8	1	4
3	7	2	8	6	4	1
3	7	4	1	8	5	2
3	7	4	2	8	1	6
3	7	4	8	2	5	1
3	7	6	2	4	1	8
3	8	1	7	6	5	2
3	8	2	7	6	1	5
3	8	4	7	1	5	2
3	8	6	5	1	4	2
3	8	7	2	6	5	1
3	8	7	5	1	4	6
3	8	7	5	4	2	1
4	1	5	8	2	7	3
4	1	7	2	6	8	3
4	1	7	5	2	8	3
4	1	7	6	3	2	8
4	1	8	5	7	3	2
4	1	8	6	2	7	5
4	1	8	6	3	7	2
4	3	1	8	7	2	6
4	3	7	2	1	8	6
4	3	8	1	7	5	2
4	3	8	2	7	1	6
4	3	8	6	2	1	5
4	6	1	3	7	2	8
4	6	1	5	2	8	3
4	6	5	1	3	2	8
4	6	5	2	1	3	8
4	6	8	3	2	7	5
4	7	1	3	6	8	2
4	7	1	5	2	8	6
4	7	1	6	2	5	8
4	7	2	1	6	8	5
4	7	2	6	3	1	8
4	7	2	8	6	3	5
4	7	5	3	2	1	8
4	8	1	5	7	2	6
4	8	2	5	7	1	3
4	8	2	6	1	7	5
4	8	5	2	1	7	6
4	8	5	2	6	1	3
4	8	5	3	1	7	2
4	8	7	2	6	5	3
4	8	7	3	6	2	5
4	8	7	5	3	2	6
5	1	4	7	2	8	3
5	1	6	2	7	3	8
5	1	8	6	3	7	4
5	3	1	6	8	2	4
5	3	1	7	2	8	6
5	3	2	7	1	8	4
5	3	6	2	1	8	4
5	3	8	1	7	4	6
5	3	8	6	2	4	1
5	3	8	7	2	4	6
5	4	2	1	8	7	3
5	4	2	7	1	3	8
5	4	6	1	3	2	8
5	4	6	2	1	3	8
5	4	8	1	3	7	6

5	4	8	2	6	3	1	7
5	4	8	3	2	7	1	6
5	7	1	3	6	8	4	2
5	7	1	6	4	3	8	2
5	7	4	1	3	8	6	2
5	7	4	2	8	3	6	1
5	7	4	3	2	8	1	6
5	7	6	2	4	3	8	1
5	7	6	3	2	4	8	1
5	8	1	6	4	7	3	2
5	8	2	6	3	7	1	4
5	8	2	7	6	4	3	1
5	8	4	3	7	2	1	6
5	8	4	6	3	2	1	7
5	8	4	7	3	2	6	1
5	8	6	2	4	7	3	1
6	1	4	7	2	8	5	3
6	1	4	7	3	8	2	5
6	1	4	8	7	3	5	2
6	1	5	2	8	7	4	3
6	1	5	7	2	4	8	3
6	1	5	8	3	7	4	2
6	1	7	5	3	8	4	2
6	3	1	7	4	8	2	5
6	3	1	8	7	5	4	2
6	3	2	8	7	4	1	5
6	3	5	1	8	4	2	7
6	3	5	2	8	1	4	7
6	3	5	8	2	4	1	7
6	3	7	1	4	8	5	2
6	3	7	2	4	8	1	5
6	3	7	2	8	5	4	1
6	4	1	3	8	7	2	5
6	4	1	5	8	3	2	7
6	4	1	7	2	5	8	3
6	4	2	1	8	7	5	3
6	4	2	7	3	1	8	5
6	4	2	8	3	7	1	5
6	4	7	1	3	5	8	2
6	5	2	1	7	4	8	3
6	5	2	8	3	7	4	1
6	5	4	1	3	8	2	7
6	5	7	2	4	3	8	1
6	5	7	3	2	4	8	1
6	8	2	5	7	4	1	3
6	8	4	2	7	5	1	3
6	8	4	3	7	1	5	2
6	8	4	7	3	5	2	1
6	8	5	2	4	7	1	3
7	1	4	6	8	2	5	3
7	1	5	8	6	4	3	2
7	1	6	5	3	8	2	4
7	3	1	6	8	5	2	4
7	3	2	6	8	1	5	4
7	3	5	2	8	1	6	4
7	3	5	8	1	4	6	2
7	3	5	8	4	2	1	6
7	3	6	8	4	2	5	1
7	3	8	6	4	1	5	2

7	4	1	5	8	2	6	3
7	4	1	8	6	5	3	2
7	4	2	5	8	1	3	6
7	4	6	2	1	5	8	3
7	4	8	2	6	5	1	3
7	4	8	3	6	1	5	2
7	4	8	5	3	1	6	2
7	5	1	3	6	8	2	4
7	5	1	6	4	2	8	3
7	5	2	6	1	4	8	3
7	5	2	6	3	1	8	4
7	5	2	8	6	4	3	1
7	5	4	2	1	8	6	3
7	5	4	3	1	8	2	6
7	5	4	8	3	2	6	1
7	5	8	3	6	4	2	1
7	6	1	5	4	8	3	2
7	6	2	5	3	8	1	4
7	6	2	5	8	4	3	1
7	6	4	2	8	5	3	1
7	6	4	3	1	8	5	2
7	6	4	3	8	2	5	1
7	6	5	3	1	4	8	2
8	1	5	7	6	2	4	3
8	1	5	7	6	3	2	4
8	1	6	5	7	2	4	3
8	1	6	5	7	3	2	4
8	3	5	7	2	1	6	4
8	3	6	1	7	5	2	4
8	3	7	6	1	5	4	2
8	3	7	6	2	5	1	4
8	3	7	6	4	2	1	5
8	4	2	7	6	1	3	5
8	4	6	3	1	7	2	5
8	4	6	5	1	3	2	7
8	4	6	5	2	1	3	7
8	4	7	1	6	5	3	2
8	4	7	3	6	2	1	5
8	4	7	5	1	3	6	2
8	5	2	6	1	7	4	3
8	5	4	1	7	2	6	3
8	5	4	6	1	3	2	7
8	5	4	6	2	1	3	7
8	5	7	3	6	2	4	1
8	6	2	5	7	3	1	4
8	6	4	1	7	5	3	2
8	6	4	3	7	2	1	5
8	6	4	7	2	5	3	1
8	6	5	1	4	7	3	2
8	6	5	3	1	7	4	2
8	6	5	3	2	7	1	4
8	7	4	6	2	5	1	3
8	7	5	3	6	1	4	2
8	7	6	5	4	3	2	1

N= 225

REFERENCES

[1] Adams,W.W. and Goldstein,L.J.. 1976. Introduction to Number Theory. Englewood Cliffs, New Jersey, Prentice-Hall.

[2] Bruck,R.H. 1955. Difference sets in a finite group. Trans. Amer. Math. Soc. 78, 464-481.

[3] Bruck,R.H. 1963. What is a loop? Studies in Modern Algebra. Edited by A.A.Albert. MAA Studies in Mathematics. Vol.2. Englewood Cliffs, New Jersey, Prentice-Hall.

[4] Doner,J.R. 1972. CIP-neofields and Combinatorial Designs. Ph.D. dissertation, The University of Michigan.

[5] Hall,M.Jr.. 1967. Combinatorial Theory. Waltham, Mass., Blaisdell.

[6] Harary,F.. 1972. Graph Theory. Reading, Mass., Addison-Wesley.

[7] Hsu,D.F.. On SIP-neofields and pure cyclic triple systems. (To appear).

[8] Hsu,D.F.. On the existence of maximal quadruple systems. (To appear).

[9] Hughes,D.R..1955. Planar division neorings. Trans. Amer. Math. Soc. 80, 502-527.

[10] Johnsen,E.C. and Storer,T.F.. 1973. Combinatorial structures in loops, I. Decomposition Theory. J. of Combinatorial Theory(A). 14, 149-166.

[11] Johnsen,E.C. and Storer,T.F.. 1974. Combinatorial structures in loops, II. Commutative Inverse Property Cyclic Neofields of Prime-Power Orders. _Pacific J. of Math_. 52, No. 1, 115-127.

[12] Johnsen,E.C. and Storer,T.F.. 1976. Combinatorial structures in loops, III. Difference Sets in Special Cyclic Neofields. _J. of Number Theory_. 18, 109-130.

[13] Johnsen,E.C. and Storer,T.F.. 1974. Combinatorial structures in loops, IV. Steiner Triple Systems in Neofields. _Math. Z._. 138, 1-14.

[14] Johnsen,E.C.. 1972. Combinatorial structures in cyclic neofields. Paper presented at the 93rd meeting of the AMS, St. Louis, Missouri, April 1.

[15] Kirkman,T.P.. 1857. On the perfect r-partition of r^2-r+1. _Trans. Historical Soc. of Lancashire and Cheshire_. 9, 127-142.

[16] Knuth,D.E.. 1965. Finite semi-fields and projective planes. _J. of Algebra_. 2, 182-217.

[17] König,D.. 1936. _Theorie der endlichen und unendlichen Gruphn_. Akad. Verlagsgesellschaft. Reprinted by Chelsea, New York, 1950.

[18] Lehmer,E.. 1953. On residue difference sets. _Canada J. Math_. 5, 425-432.

[19] Mendelsohn,N.S.. 1970. Orthogonal Steiner systems. _Aequations Math_.. 5, 268-272.

[20] Mendelsohn,N.S.. 1971. A natural generalization of Steiner triple systems. _Computers in Number Theory_.(

A.D.L. Atkin and B.J. Birch, eds.) Acad. Press., New York, 323-338.

[21] Paige,L.J.. 1949. Neofields. _Duke Math. J_..16, 39-60.

[22] Singer,J.. 1938. A theorem in finite projective geometry and some applications to number theory. _Trans. Amer. Math. Soc_.. 43, 377-385.

[23] Storer,T.F.. 1967. _Cyclotomy and Difference sets_, Chicago, Markham.

[24] Tannenbaum,P.. 1976. Abelian Steiner triple systems. _Canad. J. Math._. 28, 1251-1268.

INDEX

Vol. 670: Fonctions de Plusieurs Variables Complexes III, Proceedings, 1977. Edité par F. Norguet. XII, 394 pages. 1978.

Vol. 671: R. T. Smythe and J. C. Wierman, First-Passage Perculation on the Square Lattice. VIII, 196 pages. 1978.

Vol. 672: R. L. Taylor, Stochastic Convergence of Weighted Sums of Random Elements in Linear Spaces. VII, 216 pages. 1978.

Vol. 673: Algebraic Topology, Proceedings 1977. Edited by P. Hoffman, R. Piccinini and D. Sjerve. VI, 278 pages. 1978.

Vol. 674: Z. Fiedorowicz and S. Priddy, Homology of Classical Groups Over Finite Fields and Their Associated Infinite Loop Spaces. VI, 434 pages. 1978.

Vol. 675: J. Galambos and S. Kotz, Characterizations of Probability Distributions. VIII, 169 pages. 1978.

Vol. 676: Differential Geometrical Methods in Mathematical Physics II, Proceedings, 1977. Edited by K. Bleuler, H. R. Petry and A. Reetz. VI, 626 pages. 1978.

Vol. 677: Séminaire Bourbaki, vol. 1976/77, Exposés 489–506. IV, 264 pages. 1978.

Vol. 678: D. Dacunha-Castelle, H. Heyer et B. Roynette. Ecole d'Eté de Probabilités de Saint-Flour. VII-1977. Edité par P. L. Hennequin. IX, 379 pages. 1978.

Vol. 679: Numerical Treatment of Differential Equations in Applications, Proceedings, 1977. Edited by R. Ansorge and W. Törnig. IX, 163 pages. 1978.

Vol. 680: Mathematical Control Theory, Proceedings, 1977. Edited by W. A. Coppel. IX, 257 pages. 1978.

Vol. 681: Séminaire de Théorie du Potentiel Paris, No. 3, Directeurs: M. Brelot, G. Choquet et J. Deny. Rédacteurs: F. Hirsch et G. Mokobodzki. VII, 294 pages. 1978.

Vol. 682: G. D. James, The Representation Theory of the Symmetric Groups. V, 156 pages. 1978.

Vol. 683: Variétés Analytiques Compactes, Proceedings, 1977. Edité par Y. Hervier et A. Hirschowitz. V, 248 pages. 1978.

Vol. 684: E. E. Rosinger, Distributions and Nonlinear Partial Differential Equations. XI, 146 pages. 1978.

Vol. 685: Knot Theory, Proceedings, 1977. Edited by J. C. Hausmann. VII, 311 pages. 1978.

Vol. 686: Combinatorial Mathematics, Proceedings, 1977. Edited by D. A. Holton and J. Seberry. IX, 353 pages. 1978.

Vol. 687: Algebraic Geometry, Proceedings, 1977. Edited by L. D. Olson. V, 244 pages. 1978.

Vol. 688: J. Dydak and J. Segal, Shape Theory. VI, 150 pages. 1978.

Vol. 689: Cabal Seminar 76–77, Proceedings, 1976–77. Edited by A.S. Kechris and Y. N. Moschovakis. V, 282 pages. 1978.

Vol. 690: W. J. J. Rey, Robust Statistical Methods. VI, 128 pages. 1978.

Vol. 691: G. Viennot, Algèbres de Lie Libres et Monoïdes Libres. III, 124 pages. 1978.

Vol. 692: T. Husain and S. M. Khaleelulla, Barrelledness in Topological and Ordered Vector Spaces. IX, 258 pages. 1978.

Vol. 693: Hilbert Space Operators, Proceedings, 1977. Edited by J. M. Bachar Jr. and D. W. Hadwin. VIII, 184 pages. 1978.

Vol. 694: Séminaire Pierre Lelong – Henri Skoda (Analyse) Année 1976/77. VII, 334 pages. 1978.

Vol. 695: Measure Theory Applications to Stochastic Analysis, Proceedings, 1977. Edited by G. Kallianpur and D. Kölzow. XII, 261 pages. 1978.

Vol. 696: P. J. Feinsilver, Special Functions, Probability Semigroups, and Hamiltonian Flows. VI, 112 pages. 1978.

Vol. 697: Topics in Algebra, Proceedings, 1978. Edited by M. F. Newman. XI, 229 pages. 1978.

Vol. 698: E. Grosswald, Bessel Polynomials. XIV, 182 pages. 1978.

Vol. 699: R. E. Greene and H.-H. Wu, Function Theory on Manifolds Which Possess a Pole. III, 215 pages. 1979.

Vol. 700: Module Theory, Proceedings, 1977. Edited by C. Faith and S. Wiegand. X, 239 pages. 1979.

Vol. 701: Functional Analysis Methods in Numerical Analysis, Proceedings, 1977. Edited by M. Zuhair Nashed. VII, 333 pages. 1979.

Vol. 702: Yuri N. Bibikov, Local Theory of Nonlinear Analytic Ordinary Differential Equations. IX, 147 pages. 1979.

Vol. 703: Equadiff IV, Proceedings, 1977. Edited by J. Fábera. XIX, 441 pages. 1979.

Vol. 704: Computing Methods in Applied Sciences and Engineering, 1977, I. Proceedings, 1977. Edited by R. Glowinski and J. L. Lions. VI, 391 pages. 1979.

Vol. 705: O. Forster und K. Knorr, Konstruktion verseller Familien kompakter komplexer Räume. VII, 141 Seiten. 1979.

Vol. 706: Probability Measures on Groups, Proceedings, 1978. Edited by H. Heyer. XIII, 348 pages. 1979.

Vol. 707: R. Zielke, Discontinuous Čebyšev Systems. VI, 111 pages. 1979.

Vol. 708: J. P. Jouanolou, Equations de Pfaff algébriques. V, 255 pages. 1979.

Vol. 709: Probability in Banach Spaces II. Proceedings, 1978. Edited by A. Beck. V, 205 pages. 1979.

Vol. 710: Séminaire Bourbaki vol. 1977/78, Exposés 507–524. IV, 328 pages. 1979.

Vol. 711: Asymptotic Analysis. Edited by F. Verhulst. V, 240 pages. 1979.

Vol. 712: Equations Différentielles et Systèmes de Pfaff dans le Champ Complexe. Edité par R. Gérard et J.-P. Ramis. V, 364 pages. 1979.

Vol. 713: Séminaire de Théorie du Potentiel, Paris No. 4. Edité par F. Hirsch et G. Mokobodzki. VII, 281 pages. 1979.

Vol. 714: J. Jacod, Calcul Stochastique et Problèmes de Martingales. X, 539 pages. 1979.

Vol. 715: Inder Bir S. Passi, Group Rings and Their Augmentation Ideals. VI, 137 pages. 1979.

Vol. 716: M. A. Scheunert, The Theory of Lie Superalgebras. X, 271 pages. 1979.

Vol. 717: Grosser, Bidualräume und Vervollständigungen von Banachmoduln. III, 209 pages. 1979.

Vol. 718: J. Ferrante and C. W. Rackoff, The Computational Complexity of Logical Theories. X, 243 pages. 1979.

Vol. 719: Categorial Topology, Proceedings, 1978. Edited by H. Herrlich and G. Preuß. XII, 420 pages. 1979.

Vol. 720: E. Dubinsky, The Structure of Nuclear Fréchet Spaces. V, 187 pages. 1979.

Vol. 721: Séminaire de Probabilités XIII. Proceedings, Strasbourg, 1977/78. Edité par C. Dellacherie, P. A. Meyer et M. Weil. VII, 647 pages. 1979.

Vol. 722: Topology of Low-Dimensional Manifolds. Proceedings, 1977. Edited by R. Fenn. VI, 154 pages. 1979.

Vol. 723: W. Brandal, Commutative Rings whose Finitely Generated Modules Decompose. II, 116 pages. 1979.

Vol. 724: D. Griffeath, Additive and Cancellative Interacting Particle Systems. V, 108 pages. 1979.

Vol. 725: Algèbres d'Opérateurs. Proceedings, 1978. Edité par P. de la Harpe. VII, 309 pages. 1979.

Vol. 726: Y.-C. Wong, Schwartz Spaces, Nuclear Spaces and Tensor Products. VI, 418 pages. 1979.

Vol. 727: Y. Saito, Spectral Representations for Schrödinger Operators With Long-Range Potentials. V, 149 pages. 1979.

Vol. 728: Non-Commutative Harmonic Analysis. Proceedings, 1978. Edited by J. Carmona and M. Vergne. V, 244 pages. 1979.

Vol. 729: Ergodic Theory. Proceedings, 1978. Edited by M. Denker and K. Jacobs. XII, 209 pages. 1979.

Vol. 730: Functional Differential Equations and Approximation of Fixed Points. Proceedings, 1978. Edited by H.-O. Peitgen and H.-O. Walther. XV, 503 pages. 1979.

Vol. 731: Y. Nakagami and M. Takesaki, Duality for Crossed Products of von Neumann Algebras. IX, 139 pages. 1979.

Vol. 732: Algebraic Geometry. Proceedings, 1978. Edited by K. Lønsted. IV, 658 pages. 1979.

Vol. 733: F. Bloom, Modern Differential Geometric Techniques in the Theory of Continuous Distributions of Dislocations. XII, 206 pages. 1979.

Vol. 734: Ring Theory, Waterloo, 1978. Proceedings, 1978. Edited by D. Handelman and J. Lawrence. XI, 352 pages. 1979.

Vol. 735: B. Aupetit, Propriétés Spectrales des Algèbres de Banach. XII, 192 pages. 1979.

Vol. 736: E. Behrends, M-Structure and the Banach-Stone Theorem. X, 217 pages. 1979.

Vol. 737: Volterra Equations. Proceedings 1978. Edited by S.-O. Londen and O. J. Staffans. VIII, 314 pages. 1979.

Vol. 738: P. E. Conner, Differentiable Periodic Maps. 2nd edition, IV, 181 pages. 1979.

Vol. 739: Analyse Harmonique sur les Groupes de Lie II. Proceedings, 1976–78. Edited by P. Eymard et al. VI, 646 pages. 1979.

Vol. 740: Séminaire d'Algèbre Paul Dubreil. Proceedings, 1977–78. Edited by M.-P. Malliavin. V, 456 pages. 1979.

Vol. 741: Algebraic Topology, Waterloo 1978. Proceedings. Edited by P. Hoffman and V. Snaith. XI, 655 pages. 1979.

Vol. 742: K. Clancey, Seminormal Operators. VII, 125 pages. 1979.

Vol. 743: Romanian-Finnish Seminar on Complex Analysis. Proceedings, 1976. Edited by C. Andreian Cazacu et al. XVI, 713 pages. 1979.

Vol. 744: I. Reiner and K. W. Roggenkamp, Integral Representations. VIII, 275 pages. 1979.

Vol. 745: D. K. Haley, Equational Compactness in Rings. III, 167 pages. 1979.

Vol. 746: P. Hoffman, τ-Rings and Wreath Product Representations. V, 148 pages. 1979.

Vol. 747: Complex Analysis, Joensuu 1978. Proceedings, 1978. Edited by I. Laine, O. Lehto and T. Sorvali. XV, 450 pages. 1979.

Vol. 748: Combinatorial Mathematics VI. Proceedings, 1978. Edited by A. F. Horadam and W. D. Wallis. IX, 206 pages. 1979.

Vol. 749: V. Girault and P.-A. Raviart, Finite Element Approximation of the Navier-Stokes Equations. VII, 200 pages. 1979.

Vol. 750: J. C. Jantzen, Moduln mit einem höchsten Gewicht. III, 195 Seiten. 1979.

Vol. 751: Number Theory, Carbondale 1979. Proceedings. Edited by M. B. Nathanson. V, 342 pages. 1979.

Vol. 752: M. Barr, *-Autonomous Categories. VI, 140 pages. 1979.

Vol. 753: Applications of Sheaves. Proceedings, 1977. Edited by M. Fourman, C. Mulvey and D. Scott. XIV, 779 pages. 1979.

Vol. 754: O. A. Laudal, Formal Moduli of Algebraic Structures. III, 161 pages. 1979.

Vol. 755: Global Analysis. Proceedings, 1978. Edited by M. Grmela and J. E. Marsden. VII, 377 pages. 1979.

Vol. 756: H. O. Cordes, Elliptic Pseudo-Differential Operators – An Abstract Theory. IX, 331 pages. 1979.

Vol. 757: Smoothing Techniques for Curve Estimation. Proceedings, 1979. Edited by Th. Gasser and M. Rosenblatt. V, 245 pages. 1979.

Vol. 758: C. Năstăsescu and F. Van Oystaeyen; Graded and Filtered Rings and Modules. X, 148 pages. 1979.

Vol. 759: R. L. Epstein, Degrees of Unsolvability: Structure and Theory. XIV, 216 pages. 1979.

Vol. 760: H.-O. Georgii, Canonical Gibbs Measures. VIII, 190 pages. 1979.

Vol. 761: K. Johannson, Homotopy Equivalences of 3-Manifolds with Boundaries. 2, 303 pages. 1979.

Vol. 762: D. H. Sattinger, Group Theoretic Methods in Bifurcation Theory. V, 241 pages. 1979.

Vol. 763: Algebraic Topology, Aarhus 1978. Proceedings, 1978. Edited by J. L. Dupont and H. Madsen. VI, 695 pages. 1979.

Vol. 764: B. Srinivasan, Representations of Finite Chevalley Groups. XI, 177 pages. 1979.

Vol. 765: Padé Approximation and its Applications. Proceedings, 1979. Edited by L. Wuytack. VI, 392 pages. 1979.

Vol. 766: T. tom Dieck, Transformation Groups and Representation Theory. VIII, 309 pages. 1979.

Vol. 767: M. Namba, Families of Meromorphic Functions on Compact Riemann Surfaces. XII, 284 pages. 1979.

Vol. 768: R. S. Doran and J. Wichmann, Approximate Identities and Factorization in Banach Modules. X, 305 pages. 1979.

Vol. 769: J. Flum, M. Ziegler, Topological Model Theory. X, 151 pages. 1980.

Vol. 770: Séminaire Bourbaki vol. 1978/79 Exposés 525–542. IV, 341 pages. 1980.

Vol. 771: Approximation Methods for Navier-Stokes Problems. Proceedings, 1979. Edited by R. Rautmann. XVI, 581 pages. 1980.

Vol. 772: J. P. Levine, Algebraic Structure of Knot Modules. XI, 104 pages. 1980.

Vol. 773: Numerical Analysis. Proceedings, 1979. Edited by G. A. Watson. X, 184 pages. 1980.

Vol. 774: R. Azencott, Y. Guivarc'h, R. F. Gundy, Ecole d'Eté de Probabilités de Saint-Flour VIII-1978. Edited by P. L. Hennequin. XIII, 334 pages. 1980.

Vol. 775: Geometric Methods in Mathematical Physics. Proceedings, 1979. Edited by G. Kaiser and J. E. Marsden. VII, 257 pages. 1980.

Vol. 776: B. Gross, Arithmetic on Elliptic Curves with Complex Multiplication. V, 95 pages. 1980.

Vol. 777: Séminaire sur les Singularités des Surfaces. Proceedings, 1976-1977. Edited by M. Demazure, H. Pinkham and B. Teissier. IX, 339 pages. 1980.

Vol. 778: SK1 von Schiefkörpern. Proceedings, 1976. Edited by P. Draxl and M. Kneser. II, 124 pages. 1980.

Vol. 779: Euclidean Harmonic Analysis. Proceedings, 1979. Edited by J. J. Benedetto. III, 177 pages. 1980.

Vol. 780: L. Schwartz, Semi-Martingales sur des Variétés, et Martingales Conformes sur des Variétés Analytiques Complexes. XV, 132 pages. 1980.

Vol. 781: Harmonic Analysis Iraklion 1978. Proceedings 1978. Edited by N. Petridis, S. K. Pichorides and N. Varopoulos. V, 213 pages. 1980.

Vol. 782: Bifurcation and Nonlinear Eigenvalue Problems. Proceedings, 1978. Edited by C. Bardos, J. M. Lasry and M. Schatzman. VIII, 296 pages. 1980.

Vol. 783: A. Dinghas, Wertverteilung meromorpher Funktionen in ein- und mehrfach zusammenhängenden Gebieten. Edited by R. Nevanlinna and C. Andreian Cazacu. XIII, 145 pages. 1980.

Vol. 784: Séminaire de Probabilités XIV. Proceedings, 1978/79. Edited by J. Azéma and M. Yor. VIII, 546 pages. 1980.

Vol. 785: W. M. Schmidt, Diophantine Approximation. X, 299 pages. 1980.

Vol. 786: I. J. Maddox, Infinite Matrices of Operators. V, 122 pages. 1980.